Lectures on Harmonic Analysis

UNIVERSITY LECTURE SERIES VOLUME 29

Lectures on Harmonic Analysis

Thomas H. Wolff

Edited by
Izabella Łaba
and
Carol Shubin

AMERICAN MATHEMATICAL SOCIETY
Providence, Rhode Island

2000 *Mathematics Subject Classification.* Primary 42Bxx, 42-02; Secondary 28A75, 28A78.

For additional information and updates on this book, visit

www.ams.org/bookpages/ulect-29

Library of Congress Cataloging-in-Publication Data

Wolff, Thomas H.

    Lectures on harmonic analysis / Thomas H. Wolff; edited by Izabella Laba and Carol Shubin.

      p. cm. — (University lecture series, ISSN 1047-3998 ; v. 29)

    Includes bibliographical references.

    ISBN 0-8218-3449-5 (alk. paper)

    1. Harmonic analysis. 2. Fourier analysis. I. Laba, Izabella, 1966– II. Shubin, Carol, 1958–

III. Title. IV. University lecture series (Providence, R.I.); 29.

QA403.W65   2003

515′.2433–dc22   2003057819

Contents

Foreword   vii

Preface   ix

Chapter 1. The L^1 Fourier Transform   1

Chapter 2. The Schwartz Space   7
    Appendix: Pointwise Poincaré inequalities   11

Chapter 3. Fourier Inversion and the Plancherel Theorem   15
    Corollaries of the inversion theorem   18

Chapter 4. Some Specifics, and L^p for $p < 2$ 23

Chapter 5. The Uncertainty Principle   31

Chapter 6. The Stationary Phase Method   37

Chapter 7. The Restriction Problem   45

Chapter 8. Hausdorff Measures   57

Chapter 9. Sets with Maximal Fourier Dimension and Distance Sets   67

Chapter 10. The Kakeya Problem   79
    Bibliography   89

Chapter 11. Recent Work Connected with the Kakeya Problem   91
    List of notation   93
    11.1. The two dimensional case   93
    11.2. The higher dimensional case   101
    11.3. Circles   105
    11.4. Oscillatory integrals and Kakeya   117
    Bibliography   129

Historical Notes   133
    Bibliography   137

Foreword

It is a pleasure to introduce this testament to Tom Wolff's mathematics. Humbly and unpretentiously, Tom made repeated fundamental contributions to analysis. The core of his work deals with geometrical and measure-theoretic questions related to the Kakeya needle problem. The significance of these problems can hardly be overstated. Tom attacked them with awesome power and originality, using both geometric and combinatorial ideas, including (as far as I know) the first serious application of theoretical computer science in analysis.[1] Tom has also proven major results in other branches of analysis, particularly regarding harmonic and analytic functions. I will never forget Tom's lecture at Princeton, in which he simultaneously solved three outstanding open problems in this area, not previously known to be mutually related. Tom published few papers, but several of them went as deep as the human mind can go. "Pauca sed matura".

Charles Fefferman
Princeton, March 2003

---

[1]It should be mentioned that computer-assisted proofs play a significant and growing role in mathematics. One of Tom's best papers is, in his words, "calculator-assisted".

Preface

This book is based on a graduate course in Fourier analysis taught by Tom Wolff in the Spring of 2000 at the California Institute of Technology. Tom wrote up a set of notes which he distributed to his students and made widely available on the Internet. His intention was to publish these notes in a book format. After Tom's untimely death on July 31, 2000, I was asked to complete this work.

The selection of the material is somewhat unconventional in that the book leads us, in Tom's unique and straightforward way, through the basics directly to current research topics. Chapters 1-4 cover standard background material: the Fourier transform, convolution, the inversion theorem, the Hausdorff-Young inequality. Chapters 5 and 6 introduce the uncertainty principle and the stationary phase method. The choice of topics is highly selective, with emphasis on those frequently useful in research inspired by the problems discussed in the remaining chapters. The latter include questions related to the restriction and Kakeya conjectures, distance sets, and Fourier transforms of singular measures. These problems are diverse but often interconnected; they all combine sophisticated Fourier analysis with intriguing links to other areas of mathematics (combinatorics, number theory, partial differential equations); and they continue to stimulate first-rate work. This book focuses on laying out a solid foundation for further reading and, hopefully, research. Technicalities are kept down to the necessary minimum, and simpler but more basic methods are often favoured over the most recent ones.

The book is intended for all mathematical audiences – a novice and an expert may read it on different levels, but both should be able to find something of interest to them. A background in harmonic analysis is not necessary. Some mathematical maturity, however, will be helpful; the more junior readers should expect to work hard and to be rewarded generously for their efforts.

Tom's original manuscript constitutes Chapters 1-9 of this book. I have edited this part, clarifying a number of points and correcting typos and small errors. Most of the changes are quite minor, with the exception of Chapter 9A which was considerably expanded at the request of many readers of the original version. Chapter 10 is based on Burak Erdogan's notes of Tom's Caltech lectures; I am responsible for its final shape. The last part of

Tom's Caltech course covered the material presented in his expository article, "Recent work connected with the Kakeya problem", originally published in *Prospects in Mathematics* (H. Rossi, ed., American Mathematical Society, 1999). This article is reprinted here for the sake of completeness. I have corrected a few misprints and added footnotes (identified as "editor's notes") indicating further progress on the problems discussed; no other changes or alterations have been made.

These notes could not have been published in their present form without the help and cooperation of many people. First and foremost, I would like to thank Carol Shubin, Tom's wife and the executor of his estate, for authorizing me to edit his manuscript and for providing additional materials, including Tom's handwritten notes of a series of lectures he gave in Madison in 1996. I am grateful to Burak Erdogan for providing typeset notes which form the core of Chapter 10. Jim Colliander was kind enough to send me his notes of Tom's Madison lectures. In the Spring of 2001 I gave a series of lectures at the University of British Columbia based on Tom's manuscript; I would like to thank all those who participated, including Joel Feldman, John Fournier, Richard Froese, Ed Granirer, and Lon Rosen. Alex Iosevich, Wilhelm Schlag, and Christoph Thiele taught graduate courses based on a preliminary version of this book at the University of Missouri at Columbia, California Institute of Technology, and the University of California at Los Angeles, respectively. I would like to acknowledge the valuable comments I received from them. Michael Christ and Christopher Sogge helped me identify some of the references. I am grateful to Edward Dunne, the AMS Book Program editor, who gave his wholehearted support to this project. Finally, thanks are due to the American Mathematical Society and to the Princeton University Mathematics Department for granting us their permission to reprint Tom's expository article in this book, to Charles Fefferman who kindly provided the foreword, and to the Natural Sciences and Engineering Research Council and the National Science Foundation for their financial support.

Arguments could be made that Tom might have revised significantly the existing manuscript or included other additional topics, had he had a chance to do so. In consultation with Carol Shubin and Edward Dunne, I decided to stay as close to Tom's unfinished original as possible, preserving its character and style, and to modify and complete it only where necessary. Unfinished, perhaps, but very much alive, I hope that this book will become a lasting part of Tom's legacy.

Izabella Laba
Vancouver, March 2003

CHAPTER 1

The L^1 Fourier transform

If $f \in L^1(\mathbb{R}^n)$ then its Fourier transform is $\hat{f} : \mathbb{R}^n \to \mathbb{C}$ defined by

$$\hat{f}(\xi) = \int e^{-2\pi i x \cdot \xi} f(x) dx.$$

More generally, let $M(\mathbb{R}^n)$ be the space of finite complex-valued measures on $\mathbb{R}^n$ with the norm

$$\|\mu\| = |\mu|(\mathbb{R}^n),$$

where $|\mu|$ is the total variation. Thus $L^1(\mathbb{R}^n)$ is contained in $M(\mathbb{R}^n)$ via the identification $f \to \mu$, $d\mu = f dx$. We can generalize the definition of Fourier transform via

$$\hat{\mu}(\xi) = \int e^{-2\pi i x \cdot \xi} d\mu(x).$$

EXAMPLE 1 Let $a \in \mathbb{R}^n$ and let δ_a be the Dirac measure at a, $\delta_a(E) = 1$ if $a \in E$ and $\delta_a(E) = 0$ if $a \notin E$. Then $\hat{\delta_a}(\xi) = e^{-2\pi i a \cdot \xi}$.

EXAMPLE 2 Let $\Gamma(x) = e^{-\pi |x|^2}$. Then

(1) $$\hat{\Gamma}(\xi) = e^{-\pi |\xi|^2}.$$

PROOF. The integral in question is

$$\hat{\Gamma}(\xi) = \int e^{-2\pi i x \cdot \xi} e^{-\pi |x|^2} dx.$$

Notice that this factors as a product of one variable integrals. So it suffices to prove (1) when $n = 1$. For this we use the formula for the integral of a Gaussian: $\int_{-\infty}^{\infty} e^{-\pi x^2} dx = 1$. It follows that

$$
\begin{aligned}
\int_{-\infty}^{\infty} e^{-2\pi i x \xi} e^{-\pi x^2} dx &= \int_{-\infty}^{\infty} e^{-\pi (x + i\xi)^2} dx \cdot e^{-\pi \xi^2} \\
&= \int_{-\infty + i\xi}^{\infty + i\xi} e^{-\pi x^2} dx \cdot e^{-\pi \xi^2} \\
&= \int_{-\infty}^{\infty} e^{-\pi x^2} dx \cdot e^{-\pi \xi^2} \\
&= e^{-\pi \xi^2},
\end{aligned}
$$

where we used contour integration at the next to last line. $\square$

There are some basic estimates for the L^1 Fourier transform, which we state as Propositions 1 and 2 below. Consideration of Example 1 above shows that in complete generality not that much more can be said.

PROPOSITION 1.1. *If* $\mu \in M(\mathbb{R}^n)$ *then* $\hat{\mu}$ *is a bounded function, indeed*

$$\|\hat{\mu}\|_\infty \leq \|\mu\|_{M(\mathbb{R}^n)}. \tag{2}$$

PROOF. For any ξ,

$$
\begin{aligned}
|\hat{\mu}(\xi)| &= |\int e^{-2\pi i x \cdot \xi} d\mu(x)| \\
&\leq \int |e^{-2\pi i x \cdot \xi}| \, d|\mu|(x) \\
&= \|\mu\|.
\end{aligned}
$$

$\square$

PROPOSITION 1.2. *If* $\mu \in M(\mathbb{R}^n)$, *then* $\hat{\mu}$ *is a continuous function.*

PROOF. Fix ξ and consider

$$\hat{\mu}(\xi + h) = \int e^{-2\pi i x \cdot (\xi + h)} d\mu(x).$$

As $h \to 0$ the integrands converge pointwise to $e^{-2\pi i x \cdot \xi}$. Since all the integrands have absolute value 1 and $|\mu|(\mathbb{R}^n) < \infty$, the result follows from the dominated convergence theorem. $\square$

We now list some basic formulas for the Fourier transform; the ones listed here are roughly speaking those that do not involve any differentiations. They can all be proved by using the formula $e^{a+b} = e^a e^b$ and appropriate changes of variables. Let $f \in L^1$, $\tau \in \mathbb{R}^n$, and let T be an invertible linear map from $\mathbb{R}^n$ to $\mathbb{R}^n$.

1. Let $f_\tau(x) = f(x - \tau)$. Then

$$\hat{f_\tau}(\xi) = e^{-2\pi i \tau \cdot \xi} \hat{f}(\xi). \tag{3}$$

2. Let $e_\tau(x) = e^{2\pi i x \cdot \tau}$. Then

$$\widehat{e_\tau f}(\xi) = \hat{f}(\xi - \tau). \tag{4}$$

3. Let T^{-t} be the inverse transpose of T. Then

$$\widehat{f \circ T} = |\det(T)|^{-1} \hat{f} \circ T^{-t}. \tag{5}$$

4. Define $\tilde{f}(x) = \overline{f(-x)}$. Then

$$\widehat{\tilde{f}} = \overline{\hat{f}}. \tag{6}$$

We note some special cases of 3. If T is an orthogonal transformation (i.e. TT^t is the identity map) then $\widehat{f \circ T} = \hat{f} \circ T$, since $\det(T) = \pm 1$. In particular, this implies that if f is radial then so is $\hat{f}$, since orthogonal transformations act transitively on spheres. If T is a dilation, i.e. $Tx = r \cdot x$ for some $r > 0$, then 3. says that the Fourier transform of the function $f(rx)$ is $r^{-n} \hat{f}(r^{-1}\xi)$. Replacing r with r^{-1} and multiplying through by r^{-n}, we see that the reverse formula also holds: the Fourier transform of the function $r^{-n}f(r^{-1}x)$ is $\hat{f}(r\xi)$.

There is a general principle that if f is localized in space, then $\hat{f}$ should be smooth, and conversely if f is smooth then $\hat{f}$ should be localized. We now discuss some simple manifestations of this. Let $D(x,r) = \{y \in \mathbb{R}^n : |y - x| < r\}$.

PROPOSITION 1.3. *Suppose that $\mu \in M(\mathbb{R}^n)$ and $\operatorname{supp}\mu$ is compact. Then $\hat{\mu}$ is in C^∞ and*

$$\tag{7} D^\alpha\hat{\mu} = ((-2\pi i x)^\alpha \mu)\hat{}.$$

Furthermore, if $\operatorname{supp}\mu \subset D(0, R)$ then

$$\tag{8} \|D^\alpha\hat{\mu}\|_\infty \leq (2\pi R)^{|\alpha|}\|\mu\|.$$

We are using multiindex notation here and will do so below as well. Namely, a *multiindex* is a vector $\alpha \in \mathbb{R}^n$ whose components are nonnegative integers. If α is a multiindex then by definition

$$D^\alpha = \frac{\partial^{\alpha_1}}{\partial x_1^{\alpha_1}} \cdots \frac{\partial^{\alpha_n}}{\partial x_n^{\alpha_n}},$$

$$x^\alpha = \Pi_{j=1}^n x_j^{\alpha_j}.$$

The *length* of α, denoted $|\alpha|$, is $\sum_j \alpha_j$. One defines a partial order on multiindices via

$$\alpha \leq \beta \Leftrightarrow \alpha_i \leq \beta_i \text{ for each } i,$$

$$\alpha < \beta \Leftrightarrow \alpha \leq \beta \text{ and } \alpha \neq \beta.$$

PROOF OF PROPOSITION 1.3. Notice that (8) follows from (7) and Proposition 1 since the norm of the measure $(2\pi i x)^\alpha \mu$ is $\leq (2\pi R)^{|\alpha|}\|\mu\|$.

Furthermore, for any α the measure $(2\pi i x)^\alpha \mu$ is again a finite measure with compact support. Accordingly, if we can prove that $\hat{\mu}$ is C^1 and that (7) holds when $|\alpha| = 1$, then the lemma will follow by straightforward induction.

Fix then a value $j \in \{1, \ldots, n\}$, and let e_j be the jth standard basis vector. Also fix $\xi \in \mathbb{R}^n$, and consider the difference quotient

$$\tag{9} \Delta(h) = \frac{\hat{\mu}(\xi + he_j) - \hat{\mu}(\xi)}{h}.$$

This is equal to

$$\tag{10} \int \frac{e^{-2\pi i h x_j} - 1}{h} e^{-2\pi i \xi \cdot x} d\mu(x).$$

As $h \to 0$, the quantity

$$\frac{e^{-2\pi i h x_j} - 1}{h}$$

converges pointwise to $-2\pi i x_j$. Furthermore, $\left|\frac{e^{-2\pi i h x_j}-1}{h}\right| \leq 2\pi|x_j|$ for each h. Accordingly, the integrands in (10) are dominated by $|2\pi x_j|$, which is a bounded function on the support of μ. It follows by the dominated convergence theorem that

$$\lim_{h \to 0} \Delta(h) = \int \lim_{h \to 0} \frac{e^{-2\pi i h x_j} - 1}{h} e^{-2\pi i \xi \cdot x} d\mu(x),$$

which is equal to

$$\int -2\pi i x_j e^{-2\pi i \xi \cdot x} d\mu(x).$$

This proves the formula (7) when $|\alpha| = 1$. (7) and Proposition 2 imply that $\hat{\mu}$ is C^1. $\qquad\square$

REMARK The estimate (8) is tied to the support of μ. However, the fact that $\hat{\mu}$ is C^∞ and the formula (7) are still valid whenever μ has enough decay to justify the differentiations under the integral sign. For example, they are valid if μ has moments of all orders, i.e. $\int |x|^N d|\mu|(x) < \infty$ for all N.

The estimate (2) can be seen as justification of the idea that if μ is localized then $\hat{\mu}$ should be smooth. We now consider the converse statement, μ smooth implies $\hat{\mu}$ localized.

PROPOSITION 1.4. *Suppose that f is C^N and that $D^\alpha f \in L^1$ for all α with $0 \le |\alpha| \le N$. Then*

$$(11) \qquad \widehat{D^\alpha f}(\xi) = (2\pi i \xi)^\alpha \hat{f}(\xi)$$

when $|\alpha| \le N$ and furthermore

$$(12) \qquad |\hat{f}(\xi)| \le C(1 + |\xi|)^{-N}$$

for a suitable constant C.

The proof is based on an integration by parts which is most easily justified when f has compact support. Accordingly, we include the following lemma before giving the proof.

Let $\phi : \mathbb{R}^n \to \mathbb{R}$ be a C^∞ function with the following properties (4. is actually irrelevant for present purposes):

1. $\phi(x) = 1$ if $|x| \le 1$;
2. $\phi(x) = 0$ if $|x| \ge 2$;
3. $0 \le \phi \le 1$;
4. ϕ is radial.

Define $\phi_k(x) = \phi(\frac{x}{k})$; thus ϕ_k is similar to ϕ but lives on scale k instead of 1. If α is a multiindex, then there is a constant C_α such that $|D^\alpha \phi_k| \le \frac{C_\alpha}{k^{|\alpha|}}$ uniformly in k. Furthermore, if $\alpha \ne 0$ then the support of $D^\alpha \phi$ is contained in the region $k \le |x| \le 2k$.

LEMMA 1.5. *If f is C^N, $D^\alpha f \in L^1$ for all α with $|\alpha| \le N$ and if we let $f_k = \phi_k f$ then $\lim_{k \to \infty} \|D^\alpha f_k - D^\alpha f\|_1 = 0$ for all α with $|\alpha| \le N$.*

PROOF. It is obvious that

$$\lim_{k \to \infty} \|\phi_k D^\alpha f - D^\alpha f\|_1 = 0,$$

so it suffices to show that

$$\lim_{k \to \infty} \|D^\alpha(\phi_k f) - \phi_k D^\alpha f\|_1 = 0. \tag{13}$$

However, by the Leibniz rule

$$D^\alpha(\phi_k f) - \phi_k D^\alpha f = \sum_{0 < \beta \leq \alpha} c_\beta D^{\alpha - \beta} f D^\beta \phi_k,$$

where the c_β's are certain constants. Thus

$$\begin{aligned}
\|D^\alpha(\phi_k f) - \phi_k D^\alpha f\|_1 &\leq C \sum_{0 < \beta \leq \alpha} \|D^\beta \phi_k\|_\infty \|D^{\alpha - \beta} f\|_{L^1(\{x : |x| \geq k\})} \\
&\leq Ck^{-1} \sum_{0 < \beta \leq \alpha} \|D^{\alpha - \beta} f\|_{L^1(\{x : |x| \geq k\})}
\end{aligned}$$

The last line clearly goes to zero as $k \to \infty$. There are two reasons for this (either would suffice): the factor k^{-1}, and the fact that the L^1 norms are taken only over the region $|x| \geq k$. $\qquad\square$

PROOF OF PROPOSITION 1.4. If f is C^1 with compact support, then by integration by parts we have

$$\int \frac{\partial f}{\partial x_j}(x) e^{-2\pi i x \cdot \xi} dx = 2\pi i \xi_j \int e^{-2\pi i x \cdot \xi} f(x) dx,$$

i.e. (11) holds when $|\alpha| = 1$. An easy induction then proves (11) for all α provided that f is C^N with compact support.

To remove the compact support assumption, let f_k be as in Lemma 1.5. Then (11) holds for f_k. Now we pass to the limit as $k \to \infty$. On the one hand $\widehat{D^\alpha f_k}$ converges uniformly to $\widehat{D^\alpha f}$ as $k \to \infty$ by Lemma 1.5 and Proposition 1.1. On the other hand $\hat{f}_k$ converges uniformly to $\hat{f}$, so $(2\pi i \xi)^\alpha \hat{f}_k$ converges to $(2\pi i \xi)^\alpha \hat{f}$ pointwise. This proves (11) in general.

To prove (12), observe that (11) and Proposition 1 imply that $\xi^\alpha \hat{f} \in L^\infty$ if $|\alpha| \leq N$. On the other hand, it is easy to estimate

$$C_N^{-1}(1 + |\xi|)^N \leq \sum_{|\alpha| \leq N} |\xi^\alpha| \leq C_N(1 + |\xi|)^N, \tag{14}$$

so (12) follows. $\qquad\square$

Together with (14), let us note the inequality

$$1 + |x| \leq (1 + |y|)(1 + |x - y|), \quad x, y \in \mathbb{R}^n \tag{15}$$

which will be used several times below.

CHAPTER 2

The Schwartz Space

The Schwartz space $\mathcal{S}$ is the space of functions $f : \mathbb{R}^n \to \mathbb{C}$ such that:

(1) f is C^∞,
(2) $x^\alpha D^\beta f$ is a bounded function for each pair of multiindices α and β.

For $f \in \mathcal{S}$ we define

$$\|f\|_{\alpha\beta} = \|x^\alpha D^\beta f\|_\infty.$$

It is possible to see that $\mathcal{S}$ with the family of norms $\|\cdot\|_{\alpha\beta}$ is a Fréchet space, but we don't discuss such questions here (see [**27**]). However, we define a notion of sequential convergence in $\mathcal{S}$:

A sequence $\{f_k\} \subset \mathcal{S}$ *converges in* $\mathcal{S}$ to $f \in \mathcal{S}$ if $\lim_{k\to\infty} \|f_k - f\|_{\alpha\beta} = 0$ for each pair of multiindices α and β.

EXAMPLES: 1. Let C_0^∞ be the C^∞ functions with compact support. Then $C_0^\infty \subset \mathcal{S}$.

Namely, to prove that $x^\alpha D^\beta f$ is bounded, just note that if $f \in C_0^\infty$ then $D^\beta f$ is a continuous function with compact support, hence bounded, and that x^α is a bounded function on the support of $D^\beta f$.

2. Let $f(x) = e^{-\pi|x|^2}$. Then $f \in \mathcal{S}$.

For the proof, notice that if $p(x)$ is a polynomial, then any first partial derivative $\frac{\partial}{\partial x_j}(p(x)e^{-\pi|x|^2})$ is again of the form $q(x)e^{-\pi|x|^2}$ for some polynomial q. It follows by induction that each $D^\beta f$ is a polynomial times f for each β. Hence $x^\alpha D^\beta f$ is a polynomial times f for each α and β. This implies using L'Hospital's rule that $x^\alpha D^\beta f$ is bounded for each α and β.

3. The following functions are not in $\mathcal{S}$: $f_N(x) = (1 + |x|^2)^{-N}$ for any given N, and $g(x) = e^{-\pi|x|^2} \sin(e^{\pi|x|^2})$. Roughly, although f_N decays rapidly at ∞, it does not decay rapidly enough, whereas g decays rapidly enough but its derivatives do not decay. Detailed verification is left to reader.

We now discuss some simple properties of $\mathcal{S}$, then some which are slightly less simple.

I. $\mathcal{S}$ is closed under differentiations and under multiplication by polynomials. Furthermore, these operations are continuous on $\mathcal{S}$ in the sense that they preserve sequential convergence. Also $f, g \in \mathcal{S}$ implies $fg \in \mathcal{S}$.

PROOF. Let $f \in \mathcal{S}$. If γ is a multiindex, then $x^\alpha D^\beta (D^\gamma)f = x^\alpha D^{\beta+\gamma} f$, which is bounded since $f \in \mathcal{S}$. Hence $D^\gamma f \in \mathcal{S}$.

7

By the Leibniz rule, $x^\alpha D^\beta(x^\gamma f)$ is a finite sum of terms each of which is a constant multiple of

$$x^\alpha D^\delta(x^\gamma) D^{\beta-\delta} f$$

for some $\delta \leq \beta$. Furthermore, $D^\delta(x^\gamma)$ is a constant multiple of $x^{\gamma-\delta}$ if $\delta \leq \gamma$, and otherwise is zero. Thus $x^\alpha D^\beta(x^\gamma f)$ is a linear combination of monomials times derivatives of f, and is therefore bounded. Hence $x^\gamma f \in \mathcal{S}$.

The continuity statements follow from the proofs of the closure statements; we will normally omit these arguments. As an indication of how they are done, let us show that if γ is a multiindex then $f \to D^\gamma f$ is continuous. Suppose that $f_k \to f$ in $\mathcal{S}$. Fix a pair of multiindices α and β. Applying the definition of convergence with the multiindices α and $\beta + \gamma$, we have

$$\lim_{k \to \infty} \|x^\alpha D^{\beta+\gamma}(f_k - f)\|_\infty = 0.$$

Equivalently,

$$\lim_{k \to \infty} \|x^\alpha D^\beta(D^\gamma f_k - D^\gamma f)\|_\infty = 0$$

which says that $D^\gamma f_k$ converges to $D^\gamma f$.

The last statement (that $\mathcal{S}$ is an algebra) follows readily from the product rule and the definitions. $\square$

II. The following alternate definitions of $\mathcal{S}$ are often useful:

(16) $\qquad f \in \mathcal{S} \Leftrightarrow (1 + |x|)^N D^\beta f$ is bounded for each N and β,

(17) $\qquad f \in \mathcal{S} \Leftrightarrow \lim_{x \to \infty} x^\alpha D^\beta f = 0$ for each α and β.

Indeed, (16) follows from the definition and (14). The backward implication in (17) is trivial, while the forward implication follows by applying the definition with α replaced by appropriate larger multiindices, e.g. $\alpha + e_j$ for arbitrary $j \in \{1, \ldots, n\}$.

PROPOSITION 2.1. C_0^∞ is dense in $\mathcal{S}$, i.e. for any $f \in \mathcal{S}$ there is a sequence $\{f_k\} \subset C_0^\infty$ with $f_k \to f$ in $\mathcal{S}$.

PROOF. This is almost the same as the proof of Lemma 1.5. Namely, define ϕ_k as there and consider $f_k = \phi_k f$, which is evidently in C_0^∞. We must show that

$$x^\alpha D^\beta(\phi_k f) \to x^\alpha D^\beta f$$

uniformly as $k \to \infty$. For this, we estimate

$$\|x^\alpha D^\beta(\phi_k f) - x^\alpha D^\beta f\|_\infty \leq \|\phi_k x^\alpha D^\beta f - x^\alpha D^\beta f\|_\infty + \|x^\alpha(D^\beta(\phi_k f) - \phi_k D^\beta f)\|_\infty.$$

The first term is bounded by $\sup_{|x| \geq k} |x^\alpha D^\beta f|$ and therefore goes to zero as $k \to \infty$ by (17). The second term is estimated using the Leibniz rule by

(18) $$C \sum_{\gamma < \beta} \|x^\alpha D^\gamma f\|_\infty \|D^{\beta-\gamma}\phi_k\|_\infty.$$

Since $f \in \mathcal{S}$ and $\|D^{\beta-\gamma}\phi_k\| \leq \frac{C}{k}$, the expression (18) goes to zero as $k \to \infty$. $\square$

There is a stronger density statement which is sometimes needed. Define a C_0^∞ *tensor function* to be a function $f : \mathbb{R}^n \to \mathbb{C}$ of the form

$$f(x) = \prod_j \phi_j(x_j),$$

where each $\phi_j \in C_0^\infty(\mathbb{R})$.

PROPOSITION 2.1'. *Linear combinations of C_0^∞ tensor functions are dense in $\mathcal{S}$.*

PROOF. In view of Proposition 2.1 it suffices to show that if $f \in C_0^\infty$ then there is a sequence $\{g_k\}$ such that:

1. Each g_k is a linear combination of C_0^∞ tensor functions.

2. The supports of the g_k are contained in a fixed compact set E which is independent of k.

3. $D^\alpha g_k$ converges uniformly to $D^\alpha f$ for each α.

To construct $\{g_k\}$, we use the basic fact about Fourier series that if f is a C^∞ function in $\mathbb{R}^n$ which is 2π-periodic in each variable then f can be expanded in a series

$$f(\theta) = \sum_{\nu \in \mathbb{Z}^n} a_\nu e^{i\nu\cdot\theta}$$

where the $\{a_\nu\}$ satisfy

$$\sum_\nu (1 + |\nu|)^N |a_\nu| < \infty$$

for each N. Considering partial sums of the Fourier series, we therefore obtain a sequence of trigonometric polynomials p_k such that $D^\alpha p_k$ converges uniformly to $D^\alpha f$ for each α.

In constructing $\{g_k\}$ we can assume that $x \in \mathrm{supp} f$ implies $|x_j| \leq 1$, say, for each j - otherwise we work with $f(Rx)$ for suitable fixed R instead and undo the rescaling at the end. Let ϕ be a C_0^∞ function of one variable which is equal to 1 on $[-1, 1]$ and vanishes outside $[-2, 2]$. Let $\tilde{f}$ be the function which is equal to f on $[-\pi, \pi] \times \ldots \times [-\pi, \pi]$ and is 2π-periodic in each variable. Then we have a sequence of trigonometric polynomials p_k such that $D^\alpha p_k$ converges uniformly to $D^\alpha \tilde{f}$ for each α. Let $g_k(x) = \Pi_{j=1}^n \phi(x_j) \cdot p_k(x)$. Then g_k clearly satisfies 1. and 2, and an argument with the product rule as in Lemma 1.5 and Proposition 2.2 will show that $\{g_k\}$ satisfies 3. The proof is complete. $\qquad\square$

The next proposition is an alternate definition of $\mathcal{S}$ using L^1 instead of L^∞ norms.

PROPOSITION 2.2. *A C^∞ function f is in $\mathcal{S}$ if and only if the norms*

$$\|x^\alpha D^\beta f\|_1$$

are finite for each pair of multiindices α and β. Furthermore, a sequence $\{f_k\} \subset \mathcal{S}$ converges in $\mathcal{S}$ to $f \in \mathcal{S}$ if and only if

$$\lim_{k \to \infty} \|x^\alpha D^\beta(f_k - f)\|_1 = 0$$

for each α and β.

PROOF. We only prove the first part; the equivalence of the two notions of convergence follows from the proof and is left to the reader.

Suppose first that $f \in \mathcal{S}$. Fix α and β. Let $N = |\alpha| + n + 1$. Then we know that $(1 + |x|)^N D^\beta f$ is bounded. Accordingly,

$$\|x^\alpha D^\beta f\|_1 \ \leq \ \|(1 + |x|)^N D^\beta f\|_\infty \|x^\alpha (1 + |x|)^{-N}\|_1$$
$$< \ \infty$$

since the function $(1 + |x|)^{-n-1}$ is integrable.

For the converse, we first make a definition and state a lemma. If $f : \mathbb{R}^n \to \infty$ is C^k and if $x \in \mathbb{R}^n$ then

$$\Delta_k^f(x) \overset{def}{=} \sum_{|\alpha|=k} |D^\alpha f(x)|.$$

We denote $D(x, r) = \{y : |x-y| \leq r\}$. We also now start to use the notation $X \lesssim Y$ to mean that $X \leq CY$ where C is a fixed but unspecified constant. Unless explicitly stated otherwise, C may depend on the dimension n and various other parameters (such as exponents), but not on the functions or variables ($f, g, x, y, \dots$) involved.

LEMMA 2.2' *Suppose f is a C^∞ function. Then for any x*

$$|f(x)| \lesssim \sum_{0 \leq j \leq n+1} \|\Delta_j^f\|_{L^1(D(x,1))}.$$

This is contained in Lemma A.2 which is stated and proved at the end of the section.

To finish the proof of Proposition 2.2, we apply the preceding lemma to $D^\beta f$. This gives

$$|D^\beta f(x)| \lesssim \sum_{|\gamma| \leq |\beta|+n+1} \int_{D(x,1)} |D^\gamma f(y)| dy,$$

therefore

$$(1 + |x|)^N |D^\beta f(x)| \ \lesssim \ (1 + |x|)^N \sum_{|\gamma| \leq |\beta|+n+1} \int_{D(x,1)} |D^\gamma f(y)| dy$$
$$\lesssim \ \sum_{|\gamma| \leq |\beta|+n+1} \int_{D(x,1)} (1 + |y|)^N |D^\gamma f(y)| dy,$$

where we used the elementary inequality

$$1 + |x| \leq 2 \min_{y \in D(x,1)} (1 + |y|).$$

It follows that

$$\|(1 + |x|)^N |D^\beta f\|_\infty \lesssim \sum_{|\gamma| \leq |\beta|+n+1} \|(1 + |x|)^N D^\gamma f\|_1,$$

and then Proposition 2.2 follows from (14). $\qquad\square$

THEOREM 2.3. *If $f \in \mathcal{S}$ then $\hat{f} \in \mathcal{S}$. Furthermore, the map $f \to \hat{f}$ is continuous from $\mathcal{S}$ to $\mathcal{S}$.*

PROOF. As usual we explicitly prove only the first statement.

If $f \in \mathcal{S}$ then $f \in L^1$, so $\hat{f}$ is bounded. Thus if $f \in \mathcal{S}$, then $\widehat{D^\alpha x^\beta f}$ is bounded for any given α and β, since $D^\alpha x^\beta f$ is again in $\mathcal{S}$. However, Propositions 1.3 and 1.4 imply that

$$\widehat{D^\alpha x^\beta f}(\xi) = (2\pi i)^{|\alpha|}(-2\pi i)^{-|\beta|}\xi^\alpha D^\beta \hat{f}(\xi)$$

so $\xi^\alpha D^\beta \hat{f}$ is again bounded, which means that $\hat{f} \in \mathcal{S}$. $\qquad\square$

Appendix: Pointwise Poincaré Inequalities

This is a little more technical than the preceding and we will omit some details. We prove a frequently used pointwise estimate for a function in terms of integrals of its gradient, which plays a similar role to the mean value inequality in calculus. Then we prove a generalization involving higher derivatives which includes Lemma 2.3. Let ω be the volume of the unit ball.

LEMMA A.1 *Suppose that f is C^1. Then*

$$\left|f(x) - \frac{1}{\omega}\int_{D(x,1)} f(y)dy\right| \lesssim \int_{D(x,1)} \frac{|\nabla f(y)|}{|x-y|^{n-1}}dy.$$

PROOF. Applying the fundamental theorem of calculus to the function

$$t \to f(x + t(y-x))$$

shows that

$$|f(y) - f(x)| \leq |x-y| \int_0^1 |\nabla f(x + t(y-x))|dt.$$

Integrate this with respect to y over $D(x, 1)$ and divide by ω. Thus

$$\left|f(x) - \frac{1}{\omega}\int_{D(x,1)} f(y)dy\right| \leq \frac{1}{\omega}\int_{D(x,1)} |f(x) - f(y)|dy$$

$$\lesssim \int_{D(x,1)} |x-y| \int_0^1 |\nabla f(x + t(y-x))|dtdy$$

$$\tag{19} = \int_0^1 \int_{D(x,1)} |x-y||\nabla f(x + t(y-x))|dydt.$$

Make the change of variables $z = x + t(y - x)$, and then reverse the order of integration again. This leads to

$$(19) \quad = \int_{t=0}^{1} \int_{D(x,t)} t^{-1}|z - x||\nabla f(z)|\frac{dz}{t^n}dt$$

$$= \int_{D(x,1)} |z - x||\nabla f(z)| \int_{t=|z-x|}^{1} t^{-(n+1)}dtdz$$

$$\lesssim \int_{D(x,1)} |x - z|^{-(n-1)}|\nabla f(z)|dz$$

as claimed.

LEMMA A.2 *Suppose that f is C^k. Then*
$$(20)$$

$$|f(x)| \lesssim \sum_{0 \leq j < k} \|\Delta_j^f\|_{L^1(D(x,1))} + \begin{cases} \int_{D(x,1)} |x - y|^{-(n-k)}\Delta_k^f(y)dy \\ \qquad\qquad\qquad\qquad \text{if } 1 \leq k \leq n-1, \\ \int_{D(x,1)} \log \frac{1}{|x-y|}\Delta_n^f(y)dy \qquad \text{if } k = n, \\ \|\Delta_{n+1}^f\|_{L^1(D(x,1))} \qquad\qquad \text{if } k = n+1. \end{cases}$$

The case $k = n + 1$ is Lemma 2.2'.

PROOF. We will in fact prove that (20) holds for functions of the form $f = \sum_{j=1}^{N} |f_j|$, where f_j are C^k.

The case $k = 1$ follows immediately from Lemma A.1 applied to f_j's. To pass to general k we use induction based on the inequalities (here $a > 0, b > 0, |z - x| \leq$ constant)
$$(21)$$

$$\int_{y \in D(x,R)} |x-y|^{-(n-a)}|z-y|^{-(n-b)}dy \leq \begin{cases} C(R)|x - z|^{-(n-a-b)} & \text{if } a + b < n, \\ \log \frac{C(R)}{|z-x|} & \text{if } a + b = n, \\ C(R) & \text{if } a + b > n, \end{cases}$$

and

$$(22) \qquad \int_{y \in D(x,C)} |x - y|^{-(n-1)} \log \frac{1}{|z - y|}dy \leq C(R),$$

with $C(R)$ independent of x, z. Indeed, (21) may be proved by subdividing the region of integration in the three regimes $|y - x| \leq \frac{1}{2}|z - x|$, $|y - z| \leq \frac{1}{2}|z - x|$ and "the rest". On the first regime we have $|z - y| \geq \frac{1}{2}|z - x|$, hence the corresponding integral is bounded by

$$\int_{|y-x| \leq \frac{1}{2}|z-x|} |y - x|^{-(n-a)}|z - x|^{-(n-b)}dy \lesssim |z - x|^{-(n-a-b)}.$$

The second regime is similar. On the third regime we have $|x - y| \leq |z - y| + |x - z| \leq 3|z - y|$, so that the integral is bounded by

$$3 \int_{\frac{1}{2}|z-x| \leq |x-y| \leq R} |y - x|^{-(2n-a-b)} dy$$

and (21) follows. (22) may be proved similarly.

We now prove (20) by induction on k. We have done the case $k = 1$. Suppose that $2 \leq k \leq n - 1$ and that the cases up to and including $k - 1$ have been proved. Then

$$|f(x)| \lesssim \sum_{j \leq k-2} \|\Delta_j^f\|_{L^1(D(x,1))} + \int_{D(x,1)} |x - y|^{-(n-k+1)} \Delta_{k-1}^f(y) dy$$

$$\lesssim \sum_{j \leq k-2} \|\Delta_j^f\|_{L^1(D(x,1))} + \int_{D(x,1)} |x - y|^{-(n-k+1)} \int_{D(y,1)} \Delta_{k-1}^f(z) dz dy$$

$$+ \int_{D(x,1)} |x - y|^{-(n-k+1)} \int_{D(y,1)} |y - z|^{-(n-1)} \Delta_k^f(z) dz dy$$

$$\lesssim \sum_{j \leq k-1} \|\Delta_j^f\|_{L^1(D(x,2))} + \int_{D(x,2)} |x - z|^{-(n-k)} \Delta_k^f(z) dz.$$

For the first two inequalities we used (20) with k replaced by $k - 1$ and 1 respectively, and for the last inequality we reversed the order of integration and used (21). The disc $D(x, 2)$ can be replaced by $D(x, 1)$ using rescaling, so we have proved (20) for $k \leq n - 1$.

To pass from $k = n - 1$ to $k = n$ we argue similarly using the second case of (21), and to pass from $k = n$ to $k = n + 1$ we argue similarly using (22). $\qquad \square$

CHAPTER 3

Fourier Inversion and the Plancherel Theorem

Convolution of ϕ and f is defined as follows:

$$(23) \qquad \phi * f(x) = \int \phi(y) f(x - y) dy$$

We assume that the reader has seen this definition before but will summarize some facts, mostly without giving the proofs. Note first that convolution is commutative: the integral defining $f * \phi$ is convergent for the same values of x as (23), and $f * \phi = \phi * f$. This follows by making the change of variables $y \to x - y$. Notice also that

$$\mathrm{supp}(\phi * f) \subset \mathrm{supp}\,\phi + \mathrm{supp}\,f,$$

where the sum $E + F$ means $\{x + y : x \in E, y \in F\}$. There is an issue of the appropriate conditions on ϕ and f under which the integral (23) makes sense. We recall the following.

1. If $\phi \in L^1$ and $f \in L^p$, $1 \le p \le \infty$, then the integral (23) is an absolutely convergent Lebesgue integral for a.e. x and

$$(24) \qquad \|\phi * f\|_p \le \|\phi\|_1 \|f\|_p.$$

This is obvious when $p = \infty$, and for $p = 1$ it follows by Fubini's theorem and a change of variables. The general case can be obtained by interpolation, see the Riesz-Thorin theorem in Chapter 4.

2. If ϕ is a continuous function with compact support and $f \in L^1_{loc}$, then the integral (23) is an absolutely convergent Lebesgue integral for every x. Moreover, $\phi * f$ is continuous; this follows by rewriting (23) as $\int \phi(x - y) f(y) dy$ and applying the dominated convergence theorem.

3. If $\phi \in L^p$ and $f \in L^{p'}$, $\frac{1}{p} + \frac{1}{p'} = 1$, then by Hölder's inequality (23) is an absolutely convergent Lebesgue integral for every x, and

$$(25) \qquad \|\phi * f\|_\infty \le \|\phi\|_p \|f\|_{p'}.$$

To see that $\phi * f$ is continuous we use (25) and the fact that $\|\phi(\cdot - u) - \phi(\cdot)\|_p \to 0$ as $|u| \to 0$ for all $\phi \in L^p$, $1 \le p < \infty$.

In many applications the function ϕ is fixed and very "nice", and one considers convolution as an operator

$$f \to \phi * f.$$

LEMMA 3.1. *If $\phi \in C_0^\infty$ and $f \in L^1_{loc}$ then $\phi * f$ is C^∞ and*

$$(26) \qquad D^\alpha(\phi * f) = (D^\alpha \phi) * f.$$

15

PROOF. It is enough to prove that $\phi * f$ is C^1 and (26) holds for multiindices of length 1, since one can then use induction. Fix j and consider difference quotients

$$d(h) = \frac{1}{h}\Big((\phi * f)(x + he_j) - (\phi * f)(x)\Big).$$

Using (23) and commutativity of convolution, we can rewrite this as

$$d(h) = \int \frac{1}{h}(\phi(x + he_j - y) - \phi(x - y))f(y)dy.$$

The quotients

$$A_h(y) = \frac{1}{h}(\phi(x + he_j - y) - \phi(x - y))$$

are bounded by $\|\frac{\partial \phi}{\partial x_j}\|_\infty$ by the mean value theorem. For fixed x and for $|h| \le 1$, the support of A_h is contained in the fixed compact set $E = D(x, 1) + \operatorname{supp} \phi$. Thus the integrands $A_h f$ are dominated by the L^1 function $\|\frac{\partial \phi}{\partial x_j}\|_\infty \chi_E |f|$. The dominated convergence theorem implies

$$\lim_{h \to 0} d(h) = \int f(y) \lim_{h \to 0} A_h(y)dy = \int f(y)\frac{\partial \phi}{\partial x_j}(x - y)dy.$$

This proves (26) (when $|\alpha| = 1$). The continuity of the partials then follows from 2. above. $\qquad\square$

COROLLARY 3.1' If $f, g \in \mathcal{S}$ then $f * g \in \mathcal{S}$.

PROOF. By Lemma 3.1 it suffices to show that if $(1 + |x|)^N f(x)$ and $(1 + |x|)^N g(x)$ are bounded for every N then so is $(1 + |x|)^N f * g(x)$. This follows by writing out the definitions and using (15); the details are left to the reader. $\qquad\square$

Convolution interacts with the Fourier transform as follows: Fourier transform converts convolution to ordinary pointwise multiplication. Thus we have the following formulas:

$$(27) \qquad\qquad \widehat{f * g} = \hat{f}\hat{g}, \ f, g \in L^1,$$

$$(28) \qquad\qquad \widehat{fg} = \hat{f} * \hat{g}, \ f, g \in \mathcal{S}.$$

(27) follows from Fubini's theorem and is in many textbooks; the proof is left to the reader. (28) then follows easily from the inversion theorem, so we defer the proof until after Theorem 3.4.

Let $\phi \in \mathcal{S}$, and assume that $\int \phi = 1$. Define $\phi^\epsilon(x) = \epsilon^{-n}\phi(\epsilon^{-1}x)$. The family of functions $\{\phi^\epsilon\}$ is called an *approximate identity*. Notice that $\int \phi^\epsilon = 1$ for all ϵ. Thus one can regard the ϕ^ϵ as roughly convergent to the Dirac mass δ_0 as $\epsilon \to 0$. Indeed, the following fact is basic but quite standard; see any reasonable book on real analysis for the proof.

LEMMA 3.2. *Let $\phi \in \mathcal{S}$ and $\int \phi = 1$. Then:*

(1) *If f is a continuous function which goes to zero at ∞ then $\phi^\epsilon * f \to f$ uniformly as $\epsilon \to 0$.*

(2) *If $f \in L^p$, $1 \le p < \infty$ then $\phi^\epsilon * f \to f$ in L^p as $\epsilon \to 0$.*

Let us note the following corollary:

LEMMA 3.3. *Suppose $f \in L^1_{loc}$. Then there is a fixed sequence $\{g_k\} \subset C_0^\infty$ such that if $p \in [1, \infty)$ and $f \in L^p$, then $g_k \to f$ in L^p. If f is continuous and goes to zero at ∞, then $g_k \to f$ uniformly.*

The reason for stating the lemma in this way is that one sometimes has to deal with several notions of convergence simultaneously, e.g., L^1 and L^2 convergence, and it is convenient to be able to approximate f in both norms simultaneously.

PROOF. Let $\psi \in C_0^\infty$, $\int \psi = 1$, $\psi \ge 0$, and let ϕ be as in Lemma 1.5. Fix a sequence $\epsilon_k \downarrow 0$. Let $g_k(x) = \phi(\frac{x}{k}) \cdot (\psi^{\epsilon_k} * f)$.

If $f \in L^p$, then for large k the quantity $\|\psi^{\epsilon_k} * f\|_{L^p(|x| \ge k)}$ is bounded by $\|f\|_{L^p(|x| \ge k-1)}$ using (24) and that $\operatorname{supp} \psi^{\epsilon_k}$ is contained in $D(0,1)$. Accordingly, $\|g_k - \psi^{\epsilon_k} * f\|_p \to 0$ as $k \to \infty$. On the other hand, $\|\psi^{\epsilon_k} * f - f\|_p \to 0$ by Lemma 3.2. If f is continuous and goes to zero at ∞, then one can argue the same way using the first part of Lemma 3.2. Smoothness of g_k follows from Lemma 3.1, so the proof is complete. $\qquad\square$

THEOREM 3.4 (Fourier inversion). *Suppose that $f \in L^1$, and assume that $\hat{f}$ is also in L^1. Then for a.e. x,*

$$(29) \qquad f(x) = \int \hat{f}(\xi) e^{2\pi i \xi \cdot x} d\xi.$$

Equivalently,

$$(30) \qquad \widehat{\hat{f}}(x) = f(-x) \text{ for a.e. } x.$$

The proof uses Lemma 3.2 and also the following facts:

A. The Gaussian $\Gamma(x) = e^{-\pi|x|^2}$ satisfies $\hat{\Gamma} = \Gamma$, and therefore also satisfies (30). So at any rate there is one function f for which Theorem 3.4 is true. In fact this implies that there are many such functions. Indeed, if we form the functions

$$\Gamma_\epsilon(x) = e^{-\pi \epsilon^2 |x|^2},$$

then we have

$$(31) \qquad \hat{\Gamma}_\epsilon(\xi) = \epsilon^{-n} e^{-\pi \frac{|\xi|^2}{\epsilon^2}}.$$

Applying this again with ϵ replaced by ϵ^{-1}, one can verify that Γ_ϵ satisfies (30). See the discussion after formula (5).

B. The *duality relation* for the Fourier transform, i.e., the following lemma.

LEMMA 3.5. *Suppose that $\mu \in M(\mathbb{R}^n)$ and $\nu \in M(\mathbb{R}^n)$. Then*

$$(32) \qquad \int \hat{\mu} d\nu = \int \hat{\nu} d\mu.$$

In particular, if $f, g \in L^1$, then

$$\int \hat{f}(x)g(x)dx = \int f(x)\hat{g}(x)dx.$$

PROOF. This follows from Fubini's theorem:

$$\int \hat{\mu}d\nu = \int\int e^{-2\pi i\xi\cdot x}d\mu(x)d\nu(\xi)$$

$$= \int\int e^{-2\pi i\xi\cdot x}d\nu(\xi)d\mu(x)$$

$$= \int \hat{\nu}d\mu.$$

$\square$

PROOF OF THEOREM 3.4. Consider the integral in (29) with a damping factor included:

$$(33) \qquad I_\epsilon(x) = \int \hat{f}(\xi)e^{-\pi\epsilon^2|\xi|^2}e^{2\pi i\xi\cdot x}d\xi.$$

We evaluate the limit as $\epsilon \to 0$ in two different ways.

1. As $\epsilon \to 0$, $I_\epsilon(x) \to \int \hat{f}(\xi)e^{2\pi i\xi\cdot x}d\xi$ for each fixed x. This follows from the dominated convergence theorem, since $\hat{f} \in L^1$.

2. With x and ϵ fixed, define $g(\xi) = e^{-\pi\epsilon^2|\xi|^2}e^{2\pi i\xi\cdot x}$. Thus

$$I_\epsilon(x) = \int f(y)\hat{g}(y)dy$$

by Lemma 3.5. On the other hand, we can evaluate $\hat{g}$ using the fact that $g(\xi) = e_x(\xi)\Gamma_\epsilon(\xi)$ and (4), (31). Thus

$$\hat{g}(y) = \widehat{\Gamma_\epsilon}(y - x) = \Gamma^\epsilon(x - y),$$

where $\Gamma^\epsilon(y) = \epsilon^{-n}\Gamma(\frac{y}{\epsilon})$ is an approximate identity as in Lemma 3.2, and we have used that Γ is even.

Accordingly,

$$I_\epsilon = f * \Gamma^\epsilon,$$

and we conclude by Lemma 3.2 that

$$I_\epsilon \to f$$

in L^1 as $\epsilon \to 0$.

Summing up, we have seen that the functions I_ϵ converge pointwise to $\int \hat{f}(\xi)e^{2\pi ix\cdot\xi}d\xi$, and converge in L^1 to f. This is only possible when (29) holds.

$\square$

Corollaries of the inversion theorem

The first corollary below is not really a corollary, but a reformulation of the proof without the assumption that $\hat{f} \in L^1$. This is the form the inversion theorem takes for general f. Notice that the integrals I_ϵ are well defined for any $f \in L^1$, since the Gaussian $e^{-\pi\epsilon^2|x|^2}$ is integrable for each fixed ϵ. Corollary 3.6 is often stated as "the Fourier transform of f is Gauss-Weierstrass summable to f", and can be compared to the theorem on Cesaro summability for Fourier series.

COROLLARY 3.6. *1. Suppose $f \in L^1$ and define $I_\epsilon(x)$ via (33). Then $I_\epsilon \to f$ in L^1 as $\epsilon \to 0$.*

2. If $1 < p < \infty$ and additionally $f \in L^p$, then $I_\epsilon \to f$ in L^p as $\epsilon \to 0$. If instead f is continuous and goes to zero at ∞, then $I_\epsilon \to f$ uniformly.

PROOF. This follows from the preceding argument showing that $I_\epsilon = \Gamma^\epsilon * f$, together with Proposition 3.2. $\square$

COROLLARY 3.7. *If $f \in L^1$ and $\hat{f} = 0$, then $f = 0$.*

This is immediate from Theorem 3.4. $\square$

THEOREM 3.8. *The Fourier transform maps $\mathcal{S}$ <u>onto</u> $\mathcal{S}$.*

PROOF. Given $f \in \mathcal{S}$, let $F(x) = f(-x)$ and let $g = \hat{F}$. Then $g \in \mathcal{S}$ by Theorem 2.3, and (30) implies

$$\hat{g}(x) = \widehat{\hat{F}}(x) = F(-x) = f(x)$$

$\square$

Let us also prove formula (28). Let $f, g \in \mathcal{S}$. Then

$$\begin{aligned} \widehat{\hat{f} * \hat{g}}(-x) &= \hat{\hat{f}}(-x)\hat{\hat{g}}(-x) \\ &= f(x)g(x) \end{aligned}$$

by (27) and Theorem 3.4. Using Theorem 3.4 again, it follows that $\hat{f} * \hat{g} = \widehat{fg}$.
$\square$

THEOREM 3.9 (Plancherel Theorem, first version). *If $u, v \in \mathcal{S}$ then*

$$\text{(34)} \qquad \int \hat{u}\overline{\hat{v}} = \int u\overline{v}.$$

PROOF. By the inversion theorem,

$$\int u(x)\overline{v}(x)dx = \int \widehat{\hat{u}}(-x)\overline{v}(x)dx = \int \widehat{\hat{u}}(x)\overline{v(-x)}dx$$

i.e.,

$$\int \hat{u}\overline{\hat{v}} = \int \widehat{\hat{u}}\tilde{v}.$$

Applying the duality relation to the right-hand side we obtain

$$\int u\overline{v} = \int \widehat{\hat{u}}\tilde{v},$$

and now (34) follows from (6). $\square$

Theorem 3.9 says that the Fourier transform restricted to Schwartz functions is an isometry in the L^2 norm. Since $\mathcal{S}$ is dense in L^2 (e.g., by Lemma 3.3) this suggests a way of extending the Fourier transform to L^2.

THEOREM 3.10 (Plancherel Theorem, second version). *There is a unique bounded operator $\mathcal{F} : L^2 \to L^2$ such that $\mathcal{F}f = \hat{f}$ when $f \in \mathcal{S}$. $\mathcal{F}$ has the following additional properties:*

(1) *$\mathcal{F}$ is a unitary operator.*
(2) *$\mathcal{F}f = \hat{f}$ if $f \in L^1 \cap L^2$.*

PROOF. The existence and uniqueness statement is immediate from Theorem 3.9, as is the fact that $\|\mathcal{F}f\|_2 = \|f\|_2$. In view of this isometry property, the range of $\mathcal{F}$ must be closed, and unitarity of $\mathcal{F}$ will follow if we show that the range is dense. However, the latter statement is immediate from Theorem 3.8 and Lemma 3.3.

It remains to prove 2. For $f \in \mathcal{S}$, 2. is true by definition. Suppose now that $f \in L^1 \cap L^2$. By Lemma 3.3, there is a sequence $\{g_k\} \subset \mathcal{S}$ which converges to f both in L^1 and in L^2. By Proposition 1.1, $\widehat{g_k}$ converges to $\hat{f}$ uniformly. On the other hand, $\widehat{g_k}$ converges to $\mathcal{F}f$ in L^2 by boundedness of the operator $\mathcal{F}$. It follows that $\mathcal{F}f = \hat{f}$. $\qquad\square$

Statement 2. allows us to use the notation $\hat{f}$ for $\mathcal{F}f$ if $f \in L^2$ without any possible ambiguity. We may therefore extend the definition of the Fourier transform to $L^1 + L^2$ (in fact to σ-finite measures of the form $\mu + f\,dx$, $\mu \in M(\mathbb{R}^n), f \in L^2$) via $\widehat{f + g} = \hat{f} + \hat{g}$.

COROLLARY 3.11. *The following form of the duality relation is valid:*

$$\int \hat{\nu}\psi = \int \hat{\psi}\,d\nu, \ \psi \in \mathcal{S},$$

if $\nu = \mu + f\,dx$, $\mu \in M(\mathbb{R}^n)$, $f \in L^2$.

PROOF. We have already proved this in Lemma 3.5 if $f = 0$, so it suffices to prove it when $\mu = 0$, i.e., to show that

$$\int \hat{f}\psi = \int f\hat{\psi}$$

if $f \in L^2$, $\psi \in \mathcal{S}$. This is true by Lemma 3.5 if $f \in L^1 \cap L^2$. Therefore it is also true for $f \in L^2$, since for fixed ψ both sides depend continuously on f (in the case of the left-hand side this follows from the Plancherel theorem). $\square$

THEOREM 3.12. *If $\mu \in M(\mathbb{R}^n)$, $f \in L^2$ and*

$$\hat{f} + \hat{\mu} = 0,$$

then $\mu = -f\,dx$. In particular, if $\mu \in M(\mathbb{R}^n)$ and $\hat{\mu} \in L^2$, then μ is absolutely continuous with respect to the Lebesgue measure with an L^2 density.

PROOF. By the Riesz representation theorem for measures on compact sets, the measure $\mu + f\,dx$ will be zero provided

$$(35) \qquad\qquad \int \phi\,d\mu + \phi f\,dx = 0$$

for continuous ϕ with compact support.

If $\phi \in C_0^\infty$ then (35) follows from Corollary 3.11. In general, we choose (e.g., by Proposition 3.2) a sequence ϕ_k in C_0^∞ which converges to ϕ uniformly and in L^2. We write down (35) for the ϕ_k's and pass to the limit. This proves (35).

To prove the last statement, suppose that $\hat{\mu} \in L^2$, and choose (by Theorem 3.10) a function $g \in L^2$ with $\hat{g} = \hat{\mu}$. Then $d\mu - g dx$ has Fourier transform zero, so by the first part of the proof $d\mu = g dx$. $\qquad\square$

All the basic formulas for the L^1 Fourier transform extend to the $L^1 + L^2$-Fourier transform by approximation arguments. This was done above in the case of the duality relation. Let us note in particular that the transformation formulas in Chapter 1 extend to $L^1 + L^2$. For example, in the case of (5), one has

$$(36) \qquad \widehat{f \circ T} = |\det(T)|^{-1} \hat{f} \circ T^{-t}$$

if $f \in L^1 + L^2$. Since we already know this when $f \in L^1$, it suffices to prove it when $f \in L^2$. Choose $\{f_k\} \subset L^1 \cap L^2$, $f_k \to f$ in L^2. Composition with T is continuous on L^2, as is Fourier transform, so we can write down (36) for the $\{f_k\}$ and pass to the limit.

The $L^1 + L^2$ domain for the Fourier transform is wide enough to include many natural examples. Note in particular that $L^p \subset L^1 + L^2$ if $p \in (1, 2)$. Furthermore, certain homogeneous functions belong to $L^1 + L^2$ although none of them can belong to L^p for any fixed p. For example, $|x|^{-a}$ belongs to $L^1 + L^2$ if $\frac{n}{2} < a < n$, since it belongs to L^1 at the origin and to L^2 at infinity. However, the $L^1 + L^2$ domain is not always sufficient. The most natural way to proceed would be to develop the idea of tempered distributions, but we don't want to do this explicitly. Instead, we further broaden the definition of Fourier transform as follows: a *tempered function* is a function $f \in L^1_{loc}(\mathbb{R}^n)$ such that

$$\int (1 + |x|)^{-N} |f(x)| dx < \infty$$

for some constant N. Roughly, f has at most polynomial growth in the sense of L^1 averages.

It is clear that if f is tempered and $\phi \in \mathcal{S}$, then $\int |\phi f| < \infty$. Furthermore, the map $\phi \to \int \phi f$ is continuous on $\mathcal{S}$. It follows that $\phi * f$ is well defined if $\phi \in \mathcal{S}$ and f is tempered, and a simple estimation shows that $\phi * f$ is again tempered.

If f and g are tempered functions, we say that g *is the distributional Fourier transform of* f if

$$(37) \qquad \int g\phi = \int f\hat{\phi}$$

for all $\phi \in \mathcal{S}$. For given f, such a function g is unique using the density properties of $\mathcal{S}$ as in several previous arguments. We denote g by $\hat{f}$. Notice also that if $f \in L^1 + L^2$, then its $L^1 + L^2$-Fourier transform coincides with its distributional Fourier transform by Corollary 3.11.

All the basic formulas, in particular (3), (4), (5), (6), extend to the case of distributional Fourier transforms, e.g., if g is the distributional Fourier transform of f, then $|\det T|^{-1} \hat{f} \circ T^{-t}$ is the distributional Fourier transform of $f \circ T$. This may be seen by making appropriate changes of variable in the integrals in (37). We indicate how these arguments are carried out by proving the extended version of formula (27). Namely, if f is tempered, $\psi \in \mathcal{S}$, and f has a distributional Fourier transform, then so does $\psi * f$ and

$$\widehat{\psi * f} = \hat{\psi}\hat{f} \tag{38}$$

The proof is as follows. Let ϕ be another Schwartz function. Then

$$
\begin{aligned}
\int (\hat{\psi}\hat{f})\phi &= \int \hat{f}(\hat{\psi}\phi) \\
&= \int f\widehat{\hat{\psi}\phi} \\
&= \int f(\hat{\hat{\psi}} * \hat{\phi}) \\
&= \int f(x) \int \psi(-(x-y))\hat{\phi}(y)\,dy\,dx \\
&= \int \psi * f(y)\hat{\phi}(y).
\end{aligned}
$$

The second line followed from the definition of distributional Fourier transform, the third line from (28), and the next to last line used the inversion theorem for ψ. Comparing the above with the definition (37), we see that we have proved (38).

Let us note also that the inversion theorem is true for distributional Fourier transforms: if f is tempered and has a distributional Fourier transform $\hat{f}$, then $\hat{f}$ has the distributional Fourier transform $f(-x)$. Here is the proof. If $\phi \in \mathcal{S}$, then

$$
\begin{aligned}
\int f(-x)\phi(x)\,dx &= \int f(x)\phi(-x)\,dx \\
&= \int f(x)\hat{\hat{\phi}}(x)\,dx \\
&= \int \hat{f}(x)\hat{\phi}(x)\,dx.
\end{aligned}
$$

We used a change of variables, the inversion theorem for ϕ, and the definition (37) of $\hat{f}$. Comparing again with (37), we have the stated result. $\qquad\square$

CHAPTER 4

Some Specifics, and L^p for $p < 2$

We first discuss a couple of basic examples where the Fourier transform can be calculated, namely powers of the distance to the origin and complex Gaussians.

PROPOSITION 4.1. *Let* $h_a(x) = \frac{\gamma(a/2)}{\pi^{a/2}}|x|^{-a}$. *Then* $\widehat{h_a} = h_{n-a}$ *in the sense of* $L^1 + L^2$ *Fourier transforms if* $\frac{n}{2} < Re\,(a) < n$, *and in the sense of distributional Fourier transforms if* $0 < Re\,(a) < n$.

Here γ is the gamma function, i.e.,

$$\gamma(s) = \int_0^\infty e^{-t} t^{s-1} dt.$$

PROOF. Suppose that a is real and $\frac{n}{2} < a < n$. Then $h_a \in L^1 + L^2$. The functions of the form $f(x) = c|x|^{-a}$ with c constant may be characterized by the following two transformation properties:
1. f is radial, i.e., $f \circ \rho = f$ for all linear $\rho : \mathbb{R}^n \to \mathbb{R}^n$ with $\rho^t \rho =$ identity.
2. f is homogeneous of degree $-a$, i.e.,

$$(39) \qquad\qquad f(\epsilon x) = \epsilon^{-a} f(x)$$

for each $\epsilon > 0$.

We will use the notation

$$(40) \qquad\qquad f_\epsilon(x) = f(\epsilon x),$$

$$(41) \qquad\qquad f^\epsilon(x) = \epsilon^{-n} f(\frac{x}{\epsilon}).$$

Let $f(x) = |x|^{-a}$, $\frac{n}{2} < a < n$. Taking Fourier transforms we obtain from 1. and 2. the following (see the discussion in Chapter 1 regarding special cases of (5)): $\hat{f}$ is radial, and $\hat{f}^\epsilon = \epsilon^{-a}\hat{f}$, which is equivalent to $\hat{f}_\epsilon = \epsilon^{-(n-a)}\hat{f}$. Hence $\hat{f} = c|x|^{-(n-a)}$, and it remains to evaluate the constant c. For this we use the duality relation, taking the Schwartz function ψ to be the Gaussian Γ. Thus

$$(42) \qquad \int |x|^{-a} e^{-\pi|x|^2} dx = c \int |x|^{-(n-a)} e^{-\pi|x|^2} dx.$$

To evaluate the left hand side, change to polar coordinates and then make the change of variable $t = \pi r^2$. Thus, if σ is the area of the unit sphere, we

23

get

$$\int |x|^{-a} e^{-\pi |x|^2} dx = \sigma \int_0^\infty e^{-\pi r^2} r^{n-a} \frac{dr}{r}$$

$$= \sigma \int_0^\infty e^{-t} \left(\frac{t}{\pi}\right)^{\frac{n-a}{2}} \frac{dt}{2t}$$

$$= \frac{\sigma}{2} \pi^{-\left(\frac{n-a}{2}\right)} \gamma\left(\frac{n-a}{2}\right),$$

and similarly the right hand side of (42) is $c\frac{\sigma}{2}\pi^{-\left(\frac{a}{2}\right)}\gamma(\frac{a}{2})$. Hence

$$c = \frac{\pi^{\frac{a}{2}} \gamma\left(\frac{n-a}{2}\right)}{\pi^{\frac{n-a}{2}} \gamma\left(\frac{a}{2}\right)},$$

and the proposition is proved in the case $\frac{n}{2} < a < n$.

For the general case, fix $\phi \in \mathcal{S}$ and consider the two integrals

$$A(z) = \int h_z \hat{\phi},$$

$$B(z) = \int h_{n-z} \phi.$$

Both A and B may be seen to be analytic in z in the indicated regime: since γ is analytic, this reduces to showing that

$$\int |x|^{-z} \phi(x) dx$$

is analytic when $\phi \in \mathcal{S}$, which may be done by using the dominated convergence theorem to justify complex differentiation under the integral sign.

By Proposition 3.14, A and B agree for z in $(\frac{n}{2}, n)$. So they agree everywhere by the uniqueness theorem. This proves that the distributional Fourier transform of h_a exists and is h_{n-a}. If $\operatorname{Re} a > \frac{n}{2}$, then $h_a \in L^1 + L^2$, so that its $L^1 + L^2$ and distributional Fourier transforms coincide. $\quad\square$

Let T be an invertible $n \times n$ real symmetric matrix. The *signature* of T is the quantity $k_+ - k_-$ where k_+ and k_- are the numbers of positive and negative eigenvalues of T, counted with multiplicity. We also define

$$G_T(x) = e^{-\pi i \langle Tx, x\rangle},$$

and observe that G_T has absolute value 1 and is therefore tempered.

PROPOSITION 4.2. *Let T be an invertible $n \times n$ real symmetric matrix with signature σ. Then G_T has a distributional Fourier transform, equal to*

$$e^{-\pi i \frac{\sigma}{4}} |\det T|^{-\frac{1}{2}} G_{-T^{-1}}.$$

REMARK This can easily be generalized to complex symmetric T with nonnegative imaginary part (the latter condition is needed, else G_T is not tempered). See [**17**], Theorem 7.6.1. If $n = 1$, we do this case in the course of the proof.

PROOF. We need to show that

$$(43) \qquad \int e^{-\pi i \langle Tx,x \rangle} \hat{\phi}(x) dx = e^{\frac{-\pi i}{4} \sigma} |\det T|^{-\frac{1}{2}} \int e^{\pi i \langle T^{-1}x,x \rangle} \phi(x) dx$$

if $\phi \in \mathcal{S}$ and T is invertible real symmetric.

First consider the $n = 1$ case. Let $\sqrt{z}$ be the branch of the square root defined on the complement of the nonpositive real numbers and positive on the positive real axis. Thus $\sqrt{\pm i} = e^{\pm \frac{\pi i}{4}}$. Accordingly, (43) with $n = 1$ is equivalent to

$$(44) \qquad \int e^{-\pi z x^2} \hat{\phi}(x) dx = (\sqrt{z})^{-1} \int e^{-\pi \frac{x^2}{z}} \phi(x) dx$$

if $\phi \in \mathcal{S}$ and z is purely imaginary and non-zero. We prove this formula by analytic continuation from the real case.

Namely, if $z = 1$ then (44) is Example 2 in Chapter 1, and the case of z real and positive then follows from scaling, i.e., the fact that the Fourier transform of f_ϵ is $\hat{f}^\epsilon$, see (5). Both sides of (44) are easily seen to be analytic in z when $\operatorname{Re} z > 0$ and continuous in z when $\operatorname{Re} z \geq 0, z \neq 0$, so (44) is proved.

Now consider the $n \geq 2$ case. Observe that if (43) is true for a given T (and all ϕ), it is true also when T is replaced by UTU^{-1} for any $U \in SO(n)$. This follows from the fact that $\widehat{f \circ U} = \hat{f} \circ U$. However, since we did not give an explicit proof of the latter fact for distributional Fourier transforms, we will now exhibit the necessary calculations. Let $S = UTU^{-1}$. Thus S and T have the same determinant and the same signature. Accordingly, if (43) holds for T then

$$
\begin{aligned}
\int e^{-\pi i \langle Sx,x \rangle} \hat{\phi}(x) dx &= \int e^{-\pi i \langle TU^{-1}x, U^{-1}x \rangle} \hat{\phi}(x) dx \\
&= \int e^{-\pi i \langle Tx,x \rangle} \hat{\phi}(Ux) dx \\
&= \int e^{-\pi i \langle Tx,x \rangle} \widehat{\phi \circ U}(x) dx \\
&= e^{-\frac{\pi i}{4} \sigma} |\det T|^{-\frac{1}{2}} \int e^{\pi i \langle T^{-1}x,x \rangle} \phi \circ U(x) dx \\
&= e^{-\frac{\pi i}{4} \sigma} |\det T|^{-\frac{1}{2}} \int e^{\pi i \langle T^{-1}U^{-1}x, U^{-1}x \rangle} \phi(x) dx \\
&= e^{-\frac{\pi i}{4} \sigma} |\det S|^{-\frac{1}{2}} \int e^{\pi i \langle S^{-1}x,x \rangle} \phi(x) dx.
\end{aligned}
$$

We used that $\widehat{\phi \circ U} = \hat{\phi} \circ U$ for Schwartz functions ϕ, see the comments after formula (5) in Chapter 1.

It therefore suffices to prove (43) when T is diagonal. If T is diagonal and ϕ is a tensor function, then the integrals in (43) factor as products of one variable integrals and (43) follows immediately from (44). The general case then follows from Proposition 2.1′ and the fact that integration against a tempered function defines a continuous linear functional on $\mathcal{S}$. $\qquad\square$

We now briefly discuss the L^p Fourier transform, $1 < p < 2$. The most basic result is the Hausdorff-Young theorem, which is a formal consequence of the Plancherel theorem and Proposition 1.1 via the following.

RIESZ-THORIN INTERPOLATION THEOREM. *Let T be a linear operator with domain $L^{p_0} + L^{p_1}$, $1 \leq p_0 < p_1 \leq \infty$. Assume that $f \in L^{p_1}$ implies*

$$(45) \qquad \|Tf\|_{q_0} \leq A_0 \|f\|_{p_0}$$

$f \in L^{p_1}$ *implies*

$$(46) \qquad \|Tf\|_{q_1} \leq A_1 \|f\|_{p_1}$$

for some $1 \leq q_0, q_1 \leq \infty$. Suppose that for a certain $\theta \in (0,1)$,

$$(47) \qquad \frac{1}{p} = \frac{1-\theta}{p_0} + \frac{\theta}{p_1}$$

and

$$(48) \qquad \frac{1}{q} = \frac{1-\theta}{q_0} + \frac{\theta}{q_1}.$$

Then $f \in L^p$ implies

$$\|Tf\|_q \leq A_0^{1-\theta} A_1^{\theta} \|f\|_p.$$

For the proof see [20], [34], or numerous other textbooks.

We will adopt the convention that when indices p and p' are used we must have

$$\frac{1}{p} + \frac{1}{p'} = 1.$$

PROPOSITION 4.3 (Hausdorff-Young). *If $1 \leq p \leq 2$ then*

$$(49) \qquad \|\hat{f}\|_{p'} \leq \|f\|_p.$$

PROOF. We interpolate between the cases $p = 1$ and 2, which we already know. Namely, apply the Riesz-Thorin theorem with $p_0 = 1, q_0 = \infty, p_1 = q_1 = 2, A_0 = A_1 = 1$. The hypotheses (45) and (46) follow from Proposition 1.1 and Theorem 3.10 respectively. For given p, q, existence of $\theta \in (0,1)$ for which (47) and (48) hold is equivalent to $1 < p < 2$ and $q = p'$. The result follows. $\qquad\square$

For later reference we insert here another basic result which follows from Riesz-Thorin, although this one (in contrast to Hausdorff-Young) could also be proved by elementary manipulation of inequalities.

PROPOSITION 4.4 (Young's inequality). *Let $\phi \in L^p$, $\psi \in L^r$, where $1 \leq p, r \leq \infty$ and $\frac{1}{p} + \frac{1}{r} \geq 1$. Let $\frac{1}{q} = \frac{1}{p} - \frac{1}{r'}$. Then the integral defining $\phi * \psi$ is absolutely convergent for a.e. x and*

$$\|\phi * \psi\|_q \leq \|\phi\|_p \|\psi\|_r.$$

PROOF. View ϕ as fixed, i.e., define

$$T\psi = \phi * \psi.$$

Inequalities (24) and (25) imply that

$$T : L^1 + L^{p'} \to L^p + L^\infty$$

with

$$\|T\psi\|_p \leq \|\phi\|_p \|\psi\|_1$$
$$\|T\psi\|_\infty \leq \|\phi\|_p \|\psi\|_{p'}.$$

If $\frac{1}{q} = \frac{1}{p} - \frac{1}{r'}$ then there is $\theta \in [0, 1]$ with

$$\frac{1}{r} = \frac{1 - \theta}{1} + \frac{\theta}{p'}$$
$$\frac{1}{q} = \frac{1 - \theta}{p} + \frac{\theta}{\infty}.$$

The result now follows from Riesz-Thorin. $\qquad\square$

REMARKS 1. Unless $p = 1$ or 2, the constant 1 in the Hausdorff-Young inequality is not the best possible; indeed the best constant is found by testing the Gaussian function Γ. This is much deeper and is due to Babenko when p' is an even integer and to Beckner [1] in general. There are some related considerations in connection with Proposition 4.4, due also to Beckner.

2. Except in the case $p = 2$ the inequality (49) is not reversible, in the sense that there is no constant C such that $\|f\|_{p'} \geq \|f\|_p$ when $f \in \mathcal{S}$. Equivalently (in view of the inversion theorem) the result does not extend to the case $p > 2$. This is not at all difficult to show, but we discuss it at some length in order to illustrate a few different techniques used for constructing examples in connection with the L^p Fourier transform. Here is the most elementary argument.

EXERCISE Using translation and multiplication by characters, construct a sequence of Schwartz functions $\{\phi_n\}$ so that
1. Each ϕ_n has the same L^p norm.
2. Each $\widehat{\phi_n}$ has the same $L^{p'}$ norm.
3. The supports of the $\widehat{\phi_n}$ are disjoint.
4. The supports of the ϕ_n are "essentially disjoint" meaning that

$$\Big\| \sum_{n=1}^{N} \phi_n \Big\|_p^p \approx \sum_{n=1}^{N} \|\phi_n\|_p^p \quad (\approx N)$$

uniformly in N.

Use this to disprove the converse of Hausdorff-Young.

Here is a second argument based on Proposition 4.2. This argument can readily be adapted to show that there are functions $f \in L^p$ for any $p > 2$ which do not have a distributional Fourier transform in our sense. See [17], Theorem 7.6.6.

Take $n = 1$ and $f_\lambda(x) = \phi(x)e^{-\pi i \lambda x^2}$, where $\phi \in C_0^\infty$ is fixed. Here λ is a large positive number. Then $\|f_\lambda\|_p$ is independent of λ for any p. By the Plancherel theorem, $\|\widehat{f_\lambda}\|_2$ is also independent of λ. On the other hand, $\widehat{f_\lambda}$ is the convolution of $\hat\phi$, which is in L^1, with $(\sqrt{i\lambda})^{-1}e^{\pi i \lambda^{-1}x^2}$, which has L^∞ norm $\lambda^{-\frac{1}{2}}$. Accordingly, if $p < 2$ then

$$\|\widehat{f_\lambda}\|_{p'} \;\leq\; \|\widehat{f_\lambda}\|_2^{\frac{2}{p'}}\,\|\widehat{f_\lambda}\|_\infty^{1-\frac{2}{p'}}$$

$$\lesssim\; \lambda^{-(\frac{1}{2}-\frac{1}{p'})}.$$

Since $\|f_\lambda\|_p$ is independent of λ, this shows that when $p < 2$ there is no constant C such that $C\|f\|_{p'} \geq \|f\|_p$ for all $f \in \mathcal{S}$. $\square$

Here now is another important technique ("randomization") and a third disproof of the converse of Hausdorff-Young.

Let $\{\omega_n\}_{n=1}^N$ be independent random variables taking values ± 1 with equal probability. Denote expectation (a.k.a. integral over the probability space in question) by $\mathbb{E}$, and probability (a.k.a. measure) by Prob. Let $\{a_n\}_{n=1}^N$ be complex numbers.

PROPOSITION 4.5 (Khinchin's inequality).

$$(50) \qquad \mathbb{E}\Big(\big|\sum_{n=1}^N a_n\omega_n\big|^p\Big) \approx \Big(\sum_{n=1}^N |a_n|^2\Big)^{\frac{p}{2}}$$

for any $0 < p < \infty$, where the implicit constants depend on p only.

PROOF. Most books on probability and many analysis books give proofs. Here is the proof in the case $p > 1$. There are three steps.

(i) When $p = 2$ it is simple to see from independence that (50) is true with equality: expand out the left side and observe that the cross-terms cancel.

(ii) The upper bound. This is best obtained as a consequence of a stronger ("subgaussian") estimate. One can clearly assume the $\{a_n\}$ are real and (52) below is for real $\{a_n\}$. Let $t > 0$. We have

$$\mathbb{E}\big(e^{t\sum_n a_n\omega_n}\big) = \prod_n \mathbb{E}\big(e^{ta_n\omega_n}\big) = \prod_n \frac{1}{2}\big(e^{ta_n} + e^{-ta_n}\big),$$

where the first equality follows from independence and the fact that $e^{x+y} = e^x e^y$. Use the numerical inequality

$$(51) \qquad \frac{1}{2}\big(e^x + e^{-x}\big) \leq e^{\frac{x^2}{2}}$$

to conclude that

$$\mathbb{E}\big(e^{t\sum_n a_n\omega_n}\big) \leq e^{\frac{t^2}{2}\sum_n a_n^2},$$

therefore

$$\text{Prob}\Big(\sum_n a_n\omega_n \geq \lambda\Big) \leq e^{-t\lambda + \frac{t^2}{2}\sum_n a_n^2}$$

for any $t > 0$ and $\lambda > 0$. Taking $t = \frac{\lambda}{\sum_n a_n^2}$ gives

$$(52) \qquad \mathrm{Prob}(\sum_n a_n \omega_n \geq \lambda) \leq e^{-\frac{\lambda^2}{2\sum_n a_n^2}},$$

hence

$$\mathrm{Prob}(|\sum_n a_n \omega_n| \geq \lambda) \leq 2 e^{-\frac{\lambda^2}{2\sum_n a_n^2}}.$$

From this and the formula for the L^p norm in terms of the distribution function,

$$\mathbb{E}(|f|^p) = p \int \lambda^{p-1} \mathrm{Prob}(|f| \geq \lambda) d\lambda$$

one gets

$$\mathbb{E}(|\sum_n a_n \omega_n|^p) \leq 2p \int \lambda^{p-1} e^{-\frac{\lambda^2}{2\sum_n a_n^2}} d\lambda = 2^{2+\frac{p}{2}} p \gamma(\frac{p}{2}) (\sum_n a_n^2)^{\frac{p}{2}}.$$

This proves the upper bound.

(iii) The lower bound. This follows from (i) and (ii) by duality. Namely

$$\begin{aligned}
\sum_n |a_n|^2 &= \mathbb{E}(|\sum_n a_n \omega_n|^2) \\
&\leq \mathbb{E}(|\sum_n a_n \omega_n|^p)^{\frac{1}{p}} \mathbb{E}(|\sum_n a_n \omega_n|^{p'})^{\frac{1}{p'}} \\
&\lesssim (\sum_n |a_n|^2)^{\frac{1}{2}} \mathbb{E}(|\sum_n a_n \omega_n|^p)^{\frac{1}{p}},
\end{aligned}$$

so that

$$\mathbb{E}(|\sum_n a_n \omega_n|^p)^{\frac{1}{p}} \gtrsim (\sum_n |a_n|^2)^{\frac{1}{2}}$$

as claimed. $\qquad \square$

To apply this in connection with the converse of Hausdorff-Young, let ϕ be a C_0^∞ function, and let $\{k_j\}_{j=1}^N$ be such that the functions $\phi_j \overset{def}{=} \phi(\cdot - k_j)$ have disjoint support. Thus $\widehat{\phi_n}(\xi) = e^{2\pi i \xi \cdot k_n} \hat{\phi}(\xi)$. The L^p norm of $\sum_{n \leq N} \omega_n \phi_n$ is independent of ω in view of the disjoint supports, indeed

$$(53) \qquad \| \sum_{n \leq N} \omega_n \phi_n \|_p = C N^{\frac{1}{p}},$$

where $C = \|\phi\|_p$.

Now consider the corresponding Fourier side norms, more precisely the expectation of their p' powers:

$$(54) \qquad \mathbb{E}(\| \sum_{n \leq N} \omega_n \widehat{\phi_n} \|_{p'}^{p'}).$$

We have by Fubini's theorem

$$
\begin{aligned}
(54) \quad &= \mathbb{E}\big(\|\sum_{n \leq N} \omega_n e^{2\pi i \xi \cdot k_n} \hat{\phi}(\xi))\|_{L^{p'}(d\xi)}^{p'} \\
&= \int_{\mathbb{R}^n} |\hat{\phi}(\xi)|^{p'} \mathbb{E}\big(|\sum_{n \leq N} \omega_n e^{2\pi i \xi \cdot k_n}|^{p'}\big) \\
&\approx N^{\frac{p'}{2}},
\end{aligned}
$$

where at the last step we used Khinchin.

It follows that we can make a choice of ω so that $\|\sum_{n \leq N} \omega_n \widehat{\phi_n}\|_{p'} \lesssim N^{\frac{1}{2}}$. If $p < 2$ and if N is large, this is much smaller than the right hand side of (53), so we are done.

CHAPTER 5

The Uncertainty Principle

The uncertainty principle is[1] the heuristic statement that if a measure μ is supported on an ellipsoid E, then for many purposes $\hat{\mu}$ may be regarded as being constant on any dual ellipsoid E^*.

The simplest rigorous statement is as follows.

PROPOSITION 5.1 (L^2 Bernstein inequality). *Assume that $f \in L^2$ and $\hat{f}$ is supported in $D(0, R)$. Then f is C^∞ and there is an estimate*

$$(55) \qquad \|D^\alpha f\|_2 \leq (2\pi R)^{|\alpha|} \|f\|_2.$$

PROOF. Essentially this is an immediate consequence of the Plancherel theorem. The Fourier inversion formula

$$(56) \qquad f(x) = \int \hat{f}(\xi) e^{2\pi i x \cdot \xi} d\xi$$

is valid (in the naive sense). Namely, note that the support assumption implies that $\hat{f} \in L^1$, so that the right side is the Fourier transform of an $L^1 \cap L^2$ function. By Theorem 3.10, it is equal to f.

Proposition 1.3 applied to $\hat{f}$ now implies that f is C^∞ and that $D^\alpha f$ is obtained by differentiation under the integral sign in (56). The estimate (55) holds since

$$\|D^\alpha f\|_2 = \|\widehat{D^\alpha f}\|_2 = \|(2\pi i \xi)^\alpha \hat{f}\|_2 \leq (2\pi R)^{|\alpha|} \|\hat{f}\|_2 = (2\pi R)^{|\alpha|} \|f\|_2. \;\square$$

A corresponding statement is also true in L^p norms, but proving this and other related results needs a different argument since there is no Plancherel theorem.

LEMMA 5.2. *There is a fixed Schwartz function ϕ such that if $f \in L^1 + L^2$ and $\hat{f}$ is supported in $D(0, R)$, then*

$$f = \phi^{R^{-1}} * f.$$

PROOF. Take $\phi \in \mathcal{S}$ so that $\hat{\phi}$ is equal to 1 on $D(0, 1)$. Thus $\widehat{\phi^{R^{-1}}}(\xi) = \hat{\phi}(R^{-1}\xi)$ is equal to 1 on $D(0, R)$, so $(\phi^{R^{-1}} * f - f)\hat{\;}$ vanishes identically. Hence $\phi^{R^{-1}} * f = f$. $\qquad\square$

PROPOSITION 5.3 (Bernstein's inequality for a disc). *Suppose that $f \in L^1 + L^2$ and $\hat{f}$ is supported in $D(0, R)$. Then*

---

[1]This should be qualified by adding "as far as we are concerned". There are various more sophisticated related statements which are also called uncertainty principle; see for example [14], [15] and references there.

31

(1) *For any α and $p \in [1, \infty]$,*

$$\|D^\alpha f\|_p \leq (CR)^{|\alpha|}\|f\|_p.$$

(2) *For any $1 \leq p \leq q \leq \infty$*

$$\|f\|_q \leq CR^{n(\frac{1}{p}-\frac{1}{q})}\|f\|_p.$$

PROOF. The function $\psi = \phi^{R^{-1}}$ satisfies

$$(57) \qquad\qquad \|\psi\|_r = CR^{\frac{n}{r'}}$$

for any $r \in [1, \infty]$, where $C = \|\phi\|_r$. Also, by the chain rule

$$(58) \qquad\qquad \|\nabla\psi\|_1 = R\|\phi\|_1.$$

We know that $f = \psi * f$. In the case of first derivatives, 1. therefore follows from (57) and (24). The general case of 1. then follows by induction.

For 2., let r satisfy $\frac{1}{q} = \frac{1}{p} - \frac{1}{r'}$. Apply Young's inequality obtaining

$$\begin{aligned}
\|f\|_q &= \|\psi * f\|_q \\
&\leq \|\psi\|_r\|f\|_p \\
&\lesssim R^{\frac{n}{r'}}\|f\|_p \\
&= R^{n(\frac{1}{p}-\frac{1}{q})}\|f\|_p.
\end{aligned}$$

$\square$

We now extend the $L^p \to L^q$ bound to ellipsoids instead of balls, using change of variables. An *ellipsoid* in $\mathbb{R}^n$ is a set of the form

$$(59) \qquad E = \{x \in \mathbb{R}^n : \sum_j \frac{|(x-a) \cdot e_j|^2}{r_j^2} \leq 1\}$$

for some $a \in \mathbb{R}^n$ (called the center of E), some choice of orthonormal basis $\{e_j\}$ (the axes) and some choice of positive numbers r_j (the axis lengths). If E and E^* are two ellipsoids, then we say that E^* is *dual* to E if E^* has the same axes as E and reciprocal axis lengths, i.e., if E is given by (59) then E^* should be of the form

$$\{x \in \mathbb{R}^n : \sum_j r_j^2|(x-b) \cdot e_j|^2 \leq 1\}$$

for some choice of the center point b.

PROPOSITION 5.4 (Bernstein's inequality for an ellipsoid). *Suppose that $f \in L^1 + L^2$ and $\hat{f}$ is supported in an ellipsoid E. Then*

$$\|f\|_q \lesssim |E|^{(\frac{1}{p}-\frac{1}{q})}\|f\|_p$$

if $1 \leq p \leq q \leq \infty$.

One could similarly extend the first part of Proposition 5.3 to ellipsoids centered at the origin, but the statement is awkward since one has to weight different directions differently, so we ignore this.

PROOF. Let k be the center of E. Let T be a linear map taking the unit ball onto $E - k$. Let $S = T^{-t}$; thus $T = S^{-t}$ also. Let $f_1(x) = e^{-2\pi i k \cdot x} f(x)$ and $g = f_1 \circ S$, so that

$$
\begin{aligned}
\hat{g}(\xi) &= |\det S|^{-1} \widehat{f_1}(S^{-t}(\xi)) \\
&= |\det S|^{-1} \hat{f}(S^{-t}(\xi + k)) \\
&= |\det T| \hat{f}(T(\xi) + k)).
\end{aligned}
$$

Thus $\hat{g}$ is supported in the unit ball, so by Proposition 5.3

$$
\|g\|_q \lesssim \|g\|_p.
$$

On the other hand,

$$
\|g\|_q = |\det S|^{-\frac{1}{q}} \|f\|_q = |\det T|^{\frac{1}{q}} \|f\|_q = |E|^{\frac{1}{q}} \|f\|_q
$$

and likewise with q replaced by p. So

$$
|E|^{\frac{1}{q}} \|f\|_q \lesssim |E|^{\frac{1}{p}} \|f\|_p
$$

as claimed. $\square$

For some purposes one needs a related "pointwise statement", roughly that if $\operatorname{supp}\hat{f} \subset E$, then for any dual ellipsoid E^* the values on E^* are controlled by the average over E^*.

To formulate this precisely, let N be a large number and let $\phi(x) = (1 + |x|^2)^{-N}$. Suppose an ellipsoid R^* is given. Define $\phi_{E^*}(x) = \phi(T(x - k))$, where k is the center of E^* and T is a selfadjoint linear map taking $E^* - k$ onto the unit ball. If T_1 and T_2 are two such maps, then $T_1 \circ T_2^{-1}$ is an orthogonal transformation, so ϕ_{E^*} is well defined. Essentially, ϕ_{E^*} is roughly equal to 1 on E^* and decays rapidly as one moves away from E^*. We could also write more explicitly

$$
\phi_{E^*}(x) = \left(1 + \sum_j \frac{|(x - k) \cdot e_j|^2}{r_j^2}\right)^{-N}.
$$

PROPOSITION 5.5. *Suppose that $f \in L^1 + L^2$ and $\hat{f}$ is supported in an ellipsoid E. Then for any dual ellipsoid E^* and any $z \in E^*$,*

$$
(60) \qquad |f(z)| \leq C_N \frac{1}{|E^*|} \int |f(x)| \phi_{E^*} dx.
$$

PROOF. Assume first that E is the unit ball, and E^* is also the unit ball. Then f is the convolution of itself with a fixed Schwartz function ψ.

Accordingly

$$\begin{aligned}
|f(z)| &\leq \int |f(x)|\,|\psi(z-x)|dx \\
&\leq C_N \int |f(x)|(1+|z-x|^2)^{-N} \\
&\leq C_N \int |f(x)|(1+|x|^2)^{-N}.
\end{aligned}$$

We used the Schwartz space bounds for ψ and that $1 + |z - x|^2 \gtrsim 1 + |x|^2$ uniformly in x when $|z| \leq 1$. This proves (60) when $E = E^* =$ unit ball.

Suppose next that E is centered at zero but E and E^* are otherwise arbitrary. Let k and T be as above, and consider

$$g(x) = f(T^{-1}x + k)).$$

Its Fourier transform is supported on $T^{-1}E$, and if T maps E^* onto the unit ball, then T^{-1} maps E onto the unit ball. Accordingly,

$$|g(y)| \leq \int \phi(x)|g(x)|dx$$

if $y \in D(0,1)$, so that

$$f(T^{-1}z + k) \leq \int \phi(x)|f(T^{-1}x + k)|dx = |\det T| \int \phi_{E^*}(x)|f(x)|dx$$

by changing variables. Since $|\det T| = \frac{1}{|E^*|}$, we get (60).

If E isn't centered at zero, then we can apply the preceding with f replaced by $e^{-2\pi i k \cdot x} f(x)$ where k is the center of E. $\qquad\square$

REMARKS

1. Proposition 5.5 is an example of an estimate "with Schwartz tails". It is not possible to make the stronger conclusion that, say, $|f(x)|$ is bounded by the average of f over the double of E^* when $x \in E^*$, even in the one dimensional case with $E = E^* =$ unit interval. For this, consider a fixed Schwartz function g with $g(0) \neq 0$ whose Fourier transform is supported in the unit interval $[-1, 1]$. Consider also the functions

$$f_N(x) = (1 - \frac{x^2}{4})^N g(x).$$

Since $\hat{f}_N$ are linear combinations of $\hat{g}$ and its derivatives, they have the same support as $\hat{g}$. Moreover, they converge pointwise boundedly to zero on $[-2, 2]$, except at the origin. It follows that there can be no estimate of the value of f_N at the origin by its average over $[-2, 2]$.

2. All the estimates related to Bernstein's inequality are sharp except for the values of the constants. For example, if E is an ellipsoid, E^* a dual ellipsoid, $N < \infty$, then there is a function f with $\mathrm{supp}\hat{f} \subset E^*$ and with

$$\begin{aligned}
&(61) && \|f\|_1 \geq |E|, \\
&(62) && |f(x)| \leq C\phi_E(x),
\end{aligned}$$

where $\phi_E = \phi_E^{(N)}$ was defined above. In the case $E = E^* =$ unit ball this is obvious: take f to be any Schwartz function with Fourier support in the unit ball and with the appropriate L^1 norm. The general case then follows as above by making changes of variable.

The estimates (61) and (62) imply that $\|f\|_p \approx |E|^{\frac{1}{p}}$ for any p, so it follows that Proposition 5.4 is also sharp.

CHAPTER 6

The Stationary Phase Method

Let ϕ be a real valued C^∞ function, let a be a C_0^∞ function, and define

$$I(\lambda) = \int e^{-\pi i \lambda \phi(x)} a(x) dx.$$

Here λ is a parameter, which we always assume to be positive. The issue is the behavior of the integral $I(\lambda)$ as $\lambda \to +\infty$.

SOME GENERAL REMARKS 1. $|I(\lambda)|$ is clearly bounded by a constant depending on a only. One may expect decay as $\lambda \to \infty$, since when λ is large the integral will involve a lot of cancellation.

2. On the other hand, if ϕ is constant then $|I(\lambda)|$ is independent of λ. So one needs to put nondegeneracy hypotheses on ϕ. As it turns out, properties of a are less important. Note also that one can always cut up a with a partition of unity, which means that the question of how fast $I(\lambda)$ decays can be "localized" to a small neighborhood of a point.

3. Suppose that $\phi_1 = \phi_2 \circ G$ where G is a smooth diffeomorphism. Then

$$
\begin{aligned}
\int e^{-\pi i \lambda \phi_2(x)} a(x) dx &= \int e^{-\pi i \lambda \phi_1(G^{-1}x)} a(x) dx \\
&= \int e^{-\pi i \lambda \phi_1(y)} a(Gy) d(Gy) \\
&= \int e^{-\pi i \lambda \phi_1(y)} a(Gy) |J_G(y)| dy
\end{aligned}
$$

where J_G is the Jacobian determinant. The function $y \to a(Gy)|J_G(y)|$ is again C_0^∞, so we see that any bound for the rate of decay of $I(\lambda)$ which is independent of the choice of a will be "diffeomorphism invariant".

4. Recall from advanced calculus [23] the normal forms for a function near a regular point or a nondegenerate critical point:

STRAIGHTENING LEMMA *Suppose $\Omega \subset \mathbb{R}^n$ is open, $f : \Omega \to \mathbb{R}$ is C^∞, $p \in \Omega$ and $\nabla f(p) \neq 0$. Then there are neighborhoods U and V of 0 and p respectively and a C^∞ diffeomorphism $G : U \to V$ with $G(0) = p$ and*

$$f \circ G(x) = f(p) + x_n.$$

MORSE LEMMA *Suppose $\Omega \subset \mathbb{R}^n$ is open, $f : \Omega \to \mathbb{R}$ is C^∞, $p \in \Omega, \nabla f(p) = 0$, and suppose that the Hessian matrix $H_f(p) = \left(\frac{\partial^2 f}{\partial x_i \partial x_j}(p) \right)$ is invertible. Then, for a unique k $(=$ number of positive eigenvalues of H_f; see*

37

Lemma 6.3 below) there are neighborhoods U and V of 0 and p respectively and a C^∞ diffeomorphism $G : U \to V$ with $G(0) = p$ and

$$f \circ G(x) = f(p) + \sum_{j=1}^{k} x_j^2 - \sum_{j=k+1}^{n} x_j^2.$$

We consider now $I(\lambda)$ first when a is supported near a regular point, and then when a is supported near a nondegenerate critical point. Degenerate critical points are easy to deal with if $n = 1$, see [**33**], Chapter 8, but in higher dimensions they are much more complicated and only the two-dimensional case has been worked out, see [**36**].

PROPOSITION 6.1 (Nonstationary phase). *Suppose $\Omega \subset \mathbb{R}^n$ is open, $\phi : \Omega \to \mathbb{R}$ is C^∞, $p \in \Omega$ and $\nabla\phi(p) \neq 0$. Suppose $a \in C_0^\infty$ has its support in a sufficiently small neighborhood of p. Then*

$$\forall N \ \exists C_N : |I(\lambda)| \leq C_N \lambda^{-N},$$

and furthermore C_N depends only on bounds for finitely many derivatives of ϕ and a and a lower bound for $|\nabla\phi(p)|$ (and on N).

PROOF. The straightening lemma and the calculation in 3. above reduce this to the case $\phi(x) = x_n + c$. In this case, letting $e_n = (0, \dots, 0, 1)$ we have

$$I(\lambda) = e^{-\pi i \lambda c} \hat{a}(\frac{\lambda}{2} e_n),$$

and this has the requisite decay by Proposition 2.3. $\square$

Now we consider the nondegenerate critical point case, and as in the preceding proof we first consider the normal form.

PROPOSITION 6.2. *Let T be a real symmetric invertible matrix with signature σ, let a be C_0^∞ (or just in $\mathcal{S}$), and define*

$$I(\lambda) = \int e^{-\pi i \lambda \langle Tx, x \rangle} a(x) dx.$$

Then, for any N,

$$I(\lambda) = e^{-\pi i \frac{\sigma}{4}} |det\, T|^{-\frac{1}{2}} \lambda^{-\frac{n}{2}} \left(a(0) + \sum_{j=1}^{N} \lambda^{-j} \mathcal{D}_j a(0) + \mathcal{O}(\lambda^{-(N+1)}) \right).$$

Here $\mathcal{D}_j$ are certain explicit homogeneous constant coefficient differential operators of order $2j$, depending on T only, and the implicit constant depends only on T and on bounds for finitely many Schwartz space seminorms of a.

PROOF. Essentially this is just another way of looking at the formula for the Fourier transform of an imaginary Gaussian. By Proposition 4.2, the definition of distributional Fourier transform, and the Fourier inversion theorem for a we have

$$I(\lambda) = e^{-\pi i \frac{\sigma}{4}} \lambda^{-\frac{n}{2}} |\det T|^{-\frac{1}{2}} \int \hat{a}(-\xi) e^{\pi i \lambda^{-1} \langle T^{-1}\xi, \xi \rangle} d\xi.$$

We can replace $\hat{a}(-\xi)$ with $\hat{a}(\xi)$ by making a change of variables, since the Gaussian is even. To understand the resulting integral, use that $\lambda^{-1} \to 0$ as $\lambda \to \infty$, so the Gaussian term is approaching 1. To make this quantitative, use Taylor's theorem for e^{ix}:

$$e^{\pi i \lambda^{-1}\langle T^{-1}\xi, \xi\rangle} = \sum_{j=0}^{N} \frac{(\pi i \lambda^{-1}\langle T^{-1}\xi, \xi\rangle)^j}{j!} + \mathcal{O}(\frac{|\xi|^{2N+2}}{\lambda^{N+1}})$$

uniformly in ξ and λ. Accordingly,

$$\int \hat{a}(\xi) e^{\pi i \lambda^{-1}\langle T^{-1}\xi, \xi\rangle}\, d\xi = \int \hat{a}(\xi)\left(1 + \sum_{j=1}^{N} \frac{(\pi i \lambda^{-1}\langle T^{-1}\xi, \xi\rangle)^j}{j!}\right) d\xi$$
$$+ \mathcal{O}\left(\int |\hat{a}(\xi)| \frac{|\xi|^{2N+2}}{\lambda^{N+1}}\right).$$

Now observe that $\int \hat{a}(\xi)\, d\xi = a(0)$ by the inversion theorem, and similarly

$$\int \hat{a}(\xi) \frac{(\pi i \langle T^{-1}\xi, \xi\rangle)^j}{j!}\, d\xi$$

is the value at zero of $\mathcal{D}_j a$ for an appropriate differential operator $\mathcal{D}_j$. This gives the result, since

$$\int |\hat{a}(\xi)| |\xi|^{2N+2}\, d\xi$$

is bounded in terms of Schwartz space seminorms of $\hat{a}$, and therefore in terms of derivatives of a. $\qquad\square$

Now we consider the case of a general phase function with a nondegenerate critical point. It is clear that this should be reducible to the Gaussian case using the Morse lemma and remark 3. above. However, there is more calculation involved than in the proof of Proposition 5.1, since we need to obtain the correct form for the asymptotic expansion. We recall the following formula which follows from the chain rule:

LEMMA 6.3. *Suppose that ϕ is smooth, $\nabla\phi(p) = 0$ and G is a smooth diffeomorphism, $G(0) = p$. Then*

$$H_{\phi \circ G}(0) = DG(0)^t H_\phi(p) DG(0).$$

Thus $H_\phi(p)$ and $H_{\phi \circ G}(0)$ have the same signature and

$$\det(H_{\phi \circ G}(0)) = J_G(0)^2 \det(H_\phi(p)).$$

PROPOSITION 6.4. *Let ϕ be C^∞ and assume that $\nabla\phi(p) = 0$ and $H_\phi(p)$ is invertible. Let σ be the signature of $H_\phi(p)$, and let $\Delta = 2^{-n}|\det(H_\phi(p))|$. Let a be C_0^∞ and supported in a sufficiently small neighborhood of p. Define*

$$I(\lambda) = \int e^{-\pi i \lambda \phi(x) dx} a(x).$$

Then, for any N,

$$I(\lambda) = e^{-\pi i \lambda \phi(p)} e^{-\pi i \frac{\sigma}{4}} \Delta^{-\frac{1}{2}} \lambda^{-\frac{n}{2}} \left(a(p) + \sum_{j=1}^{N} \lambda^{-j} \mathcal{D}_j a(p) + \mathcal{O}(\lambda^{-(N+1)}) \right).$$

Here $\mathcal{D}_j$ are certain differential operators of order[1] $\leq 2j$, with coefficients depending on ϕ, and the implicit constant depends on ϕ and on bounds for finitely many derivatives of a.

PROOF. We can assume that $\phi(p) = 0$; else we replace ϕ with $\phi - \phi(p)$. Choose a C^∞ diffeomorphism G by the Morse lemma and apply remark 3. Thus

$$I_\lambda = \int e^{-\pi i \lambda \langle Ty, y \rangle} a(Gy) |J_G(y)| dy,$$

where T is a diagonal matrix with diagonal entries ± 1 and with signature σ. Also $|J_G(0)| = \Delta^{-\frac{1}{2}}$ by Lemma 6.3 and an obvious calculation of the Hessian determinant of the function $y \to \langle Ty, y \rangle$. Let $\mathcal{D}_j$ be associated to this T as in Proposition 5.2 and let $b(y) = a(Gy)|J_G(y)|$. Then

$$I(\lambda) = e^{-\pi i \frac{\sigma}{4}} \lambda^{-\frac{n}{2}} \left(b(0) + \sum_{j=1}^{N} \lambda^{-j} \mathcal{D}_j b(0) + \mathcal{O}(\lambda^{-(N+1)}) \right)$$

by Proposition 6.2. Now $b(0) = |J_G(0)| a(p) = \Delta^{-\frac{1}{2}} a(p)$, so we can write this as

$$I(\lambda) = e^{-\pi i \frac{\sigma}{4}} \Delta^{-\frac{1}{2}} \lambda^{-\frac{n}{2}} \left(a(p) + \sum_{j=1}^{N} \lambda^{-j} \Delta^{\frac{1}{2}} \mathcal{D}_j b(0) + \mathcal{O}(\lambda^{-(N+1)}) \right).$$

Further, it is clear from the chain rule and product rule that any $2j$-th order derivative of b at the origin can be expressed as a linear combination of derivatives of a at p of order $\leq 2j$ with coefficients depending on G, i.e., on ϕ. Otherwise stated, the term $\Delta^{\frac{1}{2}} \mathcal{D}_j b(0)$ can be expressed in the form $\tilde{\mathcal{D}}_j a(p)$, where $\tilde{\mathcal{D}}_j$ is a new differential operator of order $\leq 2j$ with coefficients depending on ϕ. This gives the result. $\qquad\square$

In practice, it is often more useful to have estimates for $I(\lambda)$ instead of an asymptotic expansion. Clearly an estimate $|I(\lambda)| \lesssim \lambda^{-\frac{n}{2}}$ could be derived from Proposition 6.4, but one also sometimes needs estimates for the derivatives of $I(\lambda)$ with respect to suitable parameters. For now we just consider the technically easiest case where the parameter is λ itself.

PROPOSITION 6.5. *(i) Assume that $\nabla \phi(p) \neq 0$. Then for a supported in a small neighborhood of p, $\left| \frac{d^j I(\lambda)}{d\lambda^j} \right| \leq C_{jN} \lambda^{-N}$ for any N.*

---

[1]Actually the order is exactly $2j$ but we have no need to know that.

(ii)Assume that $\nabla\phi(p) = 0$, and $H_\phi(p)$ is invertible. Then, for a supported in a small neighborhood of p,

$$\left|\frac{d^k}{d\lambda^k}(e^{\pi i\lambda\phi(p)}I(\lambda))\right| \le C_k\lambda^{-(\frac{n}{2}+k)}.$$

PROOF. We only prove (ii), since (i) follows easily from Proposition 6.1 after differentiating under the integral sign as in the proof below. For (ii) we need the following.

CLAIM. Let $\{\phi_i\}_{i=1}^M$ be real valued smooth functions and assume that $\phi_i(p) = 0$, $\nabla\phi_i(p) = 0$. Let $\Phi = \Pi_{i=1}^M\phi_i$. Then all partial derivatives of Φ of order less than $2M$ also vanish at p.

PROOF. By the product rule any partial $D^\alpha\Phi$ is a linear combination of terms of the form

$$\prod_{i=1}^M D^{\beta_i}\phi_i$$

with $\sum_i \beta_i = \alpha$. If $|\alpha| < 2M$, then some β_i must be less than 2, so by hypothesis all such terms vanish at p.

To prove the proposition, differentiate $I(\lambda)$ under the integral sign obtaining

$$\frac{d^k(e^{\pi i\lambda\phi(p)}I(\lambda))}{d\lambda^k} = (-\pi i)^k\int(\phi(x) - \phi(p))^k a(x)e^{-\pi i\lambda(\phi(x)-\phi(p))}dx.$$

Let $b(x) = (\phi(x) - \phi(p))^k a(x)$. By the above claim all partials of b of order less than $2k$ vanish at p. Now look at the expansion in Proposition 6.4 replacing a with b and setting $N = k - 1$. By the claim the terms $\mathcal{D}_j b(p)$ must vanish when $j < k$, as well as $b(p)$ itself. Hence Proposition 5.4 shows that $\frac{d^k}{d\lambda^k}(e^{\pi i\lambda\phi(p)}I(\lambda)) = \mathcal{O}(\lambda^{-(\frac{n}{2}+k)})$ as claimed. $\square$

As an application we estimate the Fourier transform of the surface measure σ on the sphere $S^{n-1} \subset \mathbb{R}^n$. For this and for other similar calculations one wants to work with an integral over a submanifold instead of over $\mathbb{R}^n$. This is not significantly different since it is always possible to work in local coordinates. However, things are easier if one uses the local coordinates as economically as possible. Recall then that if $\phi : \mathbb{R}^n \to \mathbb{R}$ is smooth and if M is a k-dimensional submanifold, $p \in M$, and if $F : U \to M$ is a local coordinate (more precisely the inverse map to one) near p, then $\phi \circ F$ will have a critical point at $F^{-1}p$ if and only if $\nabla\phi(p)$ is orthogonal to the tangent space to M at p; in particular this is independent of the choice of F.

Notice that $\hat{\sigma}$ is a radial function, because the surface measure is rotation invariant (exercise: prove this rigorously), and is smooth by Proposition 1.3. It therefore suffices to consider $\hat{\sigma}(\lambda e_n)$ where $e_n = (0, \ldots, 0, 1)$ and $\lambda > 0$.

Put local coordinates on the sphere as follows: the first "local coordinate" is the map

$$x \to (x, \sqrt{1 - |x|^2}),$$

$$\mathbb{R}^{n-1} \supset D(0, \tfrac{1}{2}) \to S^{n-1}.$$

The second is the map

$$x \to (x, -\sqrt{1-|x|^2}),$$

$$\mathbb{R}^{n-1} \supset D(0, \tfrac{1}{2}) \to S^{n-1},$$

and the remaining ones map onto sets whose closures do not contain $\{\pm e_n\}$. Let $\{q_k\}$ be a suitable partition of unity subordinate to this covering by charts. Define $\phi(x) = e_n \cdot x$, $\phi : \mathbb{R}^n \to \mathbb{R}$. Thus the gradient of ϕ is e_n and is normal to the sphere at $\pm e_n$ only.

Now

(63)

$$\hat{\sigma}(\lambda e_n) = \int e^{-2\pi i \lambda e_n \cdot x} d\sigma(x)$$

$$= \sum_{j=1}^{k} \int e^{-2\pi i \lambda e_n \cdot x} q_k(x) d\sigma(x)$$

$$= \int_{D(0,\frac{1}{2})} e^{-2\pi i \lambda \sqrt{1-|x|^2}} \frac{q_1(x)}{\sqrt{1-|x|^2}} dx + \int_{D(0,\frac{1}{2})} e^{2\pi i \lambda \sqrt{1-|x|^2}} \frac{q_2(x)}{\sqrt{1-|x|^2}} dx$$

$$+ \sum_{k \geq 3} \int e^{-2\pi i \lambda \phi_k(x)} a_k(x) dx,$$

where the dx integrals are in $\mathbb{R}^{n-1}$, and the phase functions ϕ_k for $k \geq 3$ have no critical points in the support of a_k. The Hessian of $2\sqrt{1-|x|^2}$ at the origin is -2 times the identity matrix, and in particular is invertible. It is also clear that the first and second terms are complex conjugates. We conclude from Proposition 6.5 that

$$\hat{\sigma}(\lambda e_n) = \mathrm{Re}(a(\lambda) e^{2\pi i \lambda}) + y(\lambda)$$

with

(64)
$$\frac{d^j a(\lambda)}{d\lambda^j} \lesssim \lambda^{-\frac{n-1}{2}-j},$$

(65)
$$\frac{d^j y(\lambda)}{d\lambda^j} \lesssim \lambda^{-N}$$

for any N. In fact σ is real and even and therefore $\hat{\sigma}$ must be real valued. Multiplying y by $e^{-2\pi i \lambda}$ does not affect the estimate (65), so we can absorb y into a and rewrite this as

$$\hat{\sigma}(\lambda e_n) = \mathrm{Re}(a(\lambda) e^{2\pi i \lambda}),$$

where a satisfies (64). Since $\hat{\sigma}$ is radial, we have proved the following.

COROLLARY 6.6. *The function $\hat{\sigma}$ (is a C^∞ function and) satisfies*

$$(66) \qquad \hat{\sigma}(x) = Re(a(|x|)e^{2\pi i|x|}),$$

where for large r

$$(67) \qquad |\frac{d^j a}{dr^j}| \le C_j r^{-(\frac{n-1}{2}+j)}.$$

Furthermore, looking at the first term in (64), and using the expansion of Proposition 6.4, with $N = 0$, we can obtain the leading behavior at ∞. Namely, for the first term in (64) at its critical point $x = 0$ we have a phase function $2\sqrt{1-|x|^2}$ with $\phi(0) = 2$, $\Delta = 1$ and signature $-(n-1)$, and an amplitude $q_2(x)(1-|x|^2)^{-1/2}$ which is 1 at the critical point. By Proposition 6.4 the integral is

$$e^{-2\pi i\lambda}e^{\frac{\pi i}{4}(n-1)}\lambda^{-\frac{n-1}{2}} + \mathcal{O}(\lambda^{-\frac{n+1}{2}}).$$

The second term is the complex conjugate and the others are $\mathcal{O}(\lambda^{-N})$ for any N. Hence the quantity (64) is

$$2\lambda^{-\frac{n-1}{2}}\cos(2\pi(\lambda - \frac{n-1}{8})) + \mathcal{O}(\lambda^{-\frac{n+1}{2}})$$

and we have proved

COROLLARY 6.7. *For large $|x|$*

$$\hat{\sigma}(x) = 2|x|^{-\frac{n-1}{2}}\cos(2\pi(|x| - \frac{n-1}{8})) + \mathcal{O}(|x|^{-\frac{n+1}{2}}).$$

REMARKS Of course it is possible to consider surfaces other than the sphere, see for example [17], Theorem 7.7.14. The main point in regard to the latter is that the nondegeneracy of the critical points of the phase function which arises when calculating the Fourier transform is equivalent to nonzero Gaussian curvature, so a hypersurface with nonzero Gaussian curvature everywhere behaves essentially the same as the sphere, whereas if there are flat directions the decay becomes weaker. Obtaining derivative bounds like Corollary 6.6 in the above manner requires a somewhat more complicated version of Proposition 6.5 with ϕ and a depending on an auxiliary parameter z, which we now explain without giving the proofs.

Suppose that $\phi(x, z)$ is a C^∞ function of x and z, where $x \in \mathbb{R}^n$, and $z \in \mathbb{R}^k$ should be regarded as a parameter. Assume that for a certain p and z_0 we have $\nabla_x\phi(p, z_0) = 0$ and that the matrix of second x-partials of ϕ at (p, z_0) is invertible.

1. Prove that there are neighborhoods U of z_0 and V of p and a smooth function $\kappa : U \to V$ with the following property: if $z \in V$, then $\nabla_x\phi(x, z) = 0$ if and only if $x = \kappa(z)$.

2. Let $a(x, z)$ be C_0^∞ and supported in a small enough neighborhood of (p, z_0). Define

$$I(\lambda, z) = \int e^{-\pi i\lambda\phi(x,z)}a(x, z)dx.$$

Prove the following:

$$\left| \frac{d^{j+k}\left(e^{\pi i \lambda \phi(\kappa(z),z)} I(\lambda, z)\right)}{d\lambda^j dz^k} \right| \leq C_{jk} \lambda^{-\left(\frac{n}{2}+j\right)}.$$

This is the analogue of Proposition 6.5 for general parameters.

CHAPTER 7

The Restriction Problem

Given a function $f : \mathcal{S}^{n-1} \to \mathbb{C}$, we consider the Fourier transform

$$(68) \qquad \widehat{fd\sigma}(\xi) = \int_{S^{n-1}} f(x)e^{-2\pi i x\cdot\xi}d\sigma(x).$$

If f is smooth, then one can use stationary phase to evaluate $\widehat{fd\sigma}$ to any desired degree of precision, just as with Corollary 6.6. In particular this leads to the bound

$$|\widehat{fd\sigma}(\xi)| \leq C\|f\|_{C^2}(1+|\xi|)^{-\frac{n-1}{2}}$$

say, where $\|f\|_{C^2} = \sum_{0\leq|\alpha|\leq 2}\|D^\alpha f\|_{L^\infty}$.

On the other hand there can be no similar decay estimate for functions f which are just bounded. The reason for this is that then there is no distinguished reference point in Fourier space. Thus, if we let $f_k(x) = e^{2\pi i k\cdot x}$ and set $\xi = k$, we have

$$|\widehat{f_k d\sigma}(\xi)| = \sigma(S^{n-1}) \approx 1.$$

Taking a sum of the form $f = \sum_j j^{-2} f_{k_j}$, where $|k_j| \to \infty$ sufficiently rapidly, we obtain a continuous function f such that there is no estimate

$$|\widehat{fd\sigma}(\xi)| \leq C(1+|\xi|)^{-\epsilon}$$

for any $\epsilon > 0$.

If we however consider instead L^q norms, then the issue of a distinguished origin is no longer relevant. The following is a long-standing open problem in the area.

RESTRICTION CONJECTURE (Stein) Prove that if $f \in L^\infty(S^{n-1})$ then

$$(69) \qquad \|\widehat{fd\sigma}\|_q \leq C_q\|f\|_\infty$$

for all $q > \frac{2n}{n-1}$.

The example of a constant function shows that the regime $q > \frac{2n}{n-1}$ would be best possible. Namely, Corollary 6.7 implies that $\hat\sigma \in L^q$ if and only if $q \cdot \frac{n-1}{2} > n$.

The corresponding problem for L^2 densities f was solved in the 1970's:

THEOREM 7.1 (P. Tomas-Stein). *If $f \in L^2(S^{n-1})$ then*

$$(70) \qquad \|\widehat{fd\sigma}\|_q \leq C\|f\|_{L^2(S^{n-1})}$$

for $q \geq \frac{2n+2}{n-1}$, and this range of q is best possible.

REMARKS 1. Notice that the assumptions on q in (70) and (69) are of the form $q > q_0$ or $q \geq q_0$. The reason for this is that there is an obvious estimate

$$\|\widehat{f d\sigma}\|_\infty \leq \|f\|_1$$

by Proposition 1.1, and it follows by the Riesz-Thorin theorem that if (69) or (70) holds for a given q, then it also holds for any larger q.

2. The restriction conjecture (69) is known to be true when $n = 2$; this is due to C. Fefferman and Stein, early 1970's. See [12] and [33].

3. Of course there is a difference in the L^q exponent in (70) and the one which is conjectured for L^∞ densities. Until fairly recently it was unknown (in three or more dimensions) whether the estimate (69) was true even for some q less than the Stein-Tomas exponent $\frac{2n+2}{n-1}$. This was first shown by Bourgain [3], a paper which has been the starting point for a lot of recent work.

4. The fact that $q \geq \frac{2n+2}{n-1}$ is best possible for (70) is due I believe to A. Knapp. We now discuss the construction. Notice that in order to distinguish between L^2 and L^∞ norms, one should use a function f which is highly localized. Next, in view of the nice behavior of rectangles under the Fourier transform discussed e.g. in our Chapter 5, it is natural to take the support of f to be the intersection of S^{n-1} with a small rectangle. Now we set up the proof.

Let

$$C_\delta = \{x \in S^{n-1} : 1 - x \cdot e_n \leq \delta^2\},$$

where $e_n = (0, \ldots, 0, 1)$. Since $|x - e_n|^2 = 2(1 - x \cdot e_n)$, it is easy to show that

$$(71) \qquad |x - e_n| \leq C^{-1}\delta \Rightarrow x \in C_\delta \Rightarrow |x - e_n| \leq C\delta$$

for an appropriate constant C. Now let $f = f_\delta$ be the indicator function of C_δ. We calculate $\|f\|_{L^2(S^{n-1})}$ and $\|\widehat{f d\sigma}\|_q$. All constants are of course independent of δ.

In the first place, $\|f\|_2$ is the square root of the measure of C_δ, so by (71) and the dimensionality of the sphere we have

$$(72) \qquad \|f\|_2 \approx \delta^{\frac{n-1}{2}}.$$

The support of $f d\sigma$ is contained in the rectangle centered at e_n with length about δ^2 in the e_n direction and length about δ in the orthogonal directions. We look at $\widehat{f d\sigma}$ on the dual rectangle centered at 0. Suppose then that $|\xi_n| \leq C_1^{-1}\delta^{-2}$ and that $|\xi_j| \leq C_1^{-1}\delta^{-1}$ when $j < n$; here C_1 is a large constant. Then

$$\begin{aligned}
|\widehat{fd\sigma}(\xi)| &= \left| \int_{C_\delta} e^{-2\pi i x \cdot \xi} d\sigma(x) \right| \\
&= \left| \int_{C_\delta} e^{-2\pi i (x - e_n) \cdot \xi} d\sigma(x) \right| \\
&\geq \int_{C_\delta} \cos(2\pi (x - e_n) \cdot \xi) d\sigma(x).
\end{aligned}$$

Our conditions on ξ imply if C_1 is large enough that $|(x - e_n) \cdot \xi| \leq \frac{\pi}{3}$, say, for all $x \in C_\delta$. Accordingly,

$$|\widehat{fd\sigma}(\xi)| \geq \frac{1}{2}|C_\delta| \approx \delta^{n-1}.$$

Our set of ξ has volume about $\delta^{-(n+1)}$, so we conclude that

$$\|\widehat{fd\sigma}\|_q \gtrsim \delta^{n-1-\frac{n+1}{q}}.$$

Comparing this estimate with (72) we find that if (70) holds then

$$\delta^{n-1-\frac{n+1}{q}} \lesssim \delta^{\frac{n-1}{2}}$$

uniformly in $\delta \in (0, 1]$. Hence $n - 1 - \frac{n+1}{q} \geq \frac{n-1}{2}$, i.e. $q \geq \frac{2n+2}{n-1}$.

For future reference we record the following variant on the above example: if f is as above and $g = e^{2\pi i x \cdot \eta} T f$, where $\eta \in \mathbb{R}^n$ and T is a rotation mapping e_n to $v \in S^{n-1}$, then g is supported on

$$\{x \in S^{n-1} : 1 - x \cdot v \leq \delta^2\},$$

and

$$|\widehat{gd\sigma}| \gtrsim \delta^{n-1}$$

on a cylinder of length $C_1^{-1}\delta^{-2}$ and cross-section radius $C_1^{-1}\delta^{-1}$, centered at η and with the axis parallel to v.

Before giving the proof of Theorem 7.1 we need to discuss convolution of a Schwartz function with a measure, since this wasn't previously considered. Let $\mu \in M(\mathbb{R}^n)$; assume μ has compact support for simplicity, although this assumption is not really needed. Define

$$\phi * \mu(x) = \int \phi(x - y) d\mu(y).$$

Observe that $\phi * \mu$ is C^∞, since differentiation under the integral sign is justified as in Lemma 3.1.

It is convenient to use the notation $\check{\mu}$ for $\hat{\mu}(-x)$. We need to extend some of our formulas to the present context. In particular the following extends (28), since if $\mu \in \mathcal{S}$ then the Fourier transform of $\check{\mu}$ is μ by Theorem 3.4:

$$(73) \qquad\qquad \widehat{\phi\check{\mu}} = \hat{\phi} * \mu \text{ when } \phi \in \mathcal{S},$$

$$(74) \qquad\qquad \widehat{\phi\mu} = \hat{\phi} * \hat{\mu} \text{ when } \phi \in \mathcal{S}.$$

Notice that (73) can be interpreted naively: Proposition 1.3 and the product rule imply that $\phi\check{\mu}$ is a Schwartz function. To prove (73), by uniqueness of distributional Fourier transforms it suffices to show that

$$\int \hat{\psi}\phi\check{\mu} = \int \psi(\hat{\phi} * \mu)$$

if ψ is another Schwartz function. This is done as follows. Denote $Tx = -x$, then

$$\begin{aligned}
\int \hat{\psi}(x)\phi(x)\hat{\mu}(-x)dx &= \int \hat{\psi}(-x)\phi(-x)\hat{\mu}(x)dx \\
&= \int ((\hat{\psi}\phi) \circ T)\hat{\mu} \\
&= \int ((\hat{\psi}\phi) \circ T)\hat{\ } d\mu \text{ by the duality relation} \\
&= \int \widehat{(\hat{\psi} \circ T)} * \widehat{(\phi \circ T)} d\mu \\
&= \int \psi * (\hat{\phi} \circ T) d\mu \\
&= \int\int \psi(y)\hat{\phi}(-x + y)dy d\mu(x) \\
&= \int \psi(y)\hat{\phi} * \mu(y) dy.
\end{aligned}$$

For (74), again let ψ be another Schwartz function. Then

$$\begin{aligned}
\int \widehat{\phi\mu}\psi dx &= \int \hat{\psi}\phi d\mu \text{ by the duality relation} \\
&= \int \widehat{\check{\phi} * \psi} d\mu \\
&= \int (\check{\phi} * \psi)\hat{\mu} dx \text{ by the duality relation} \\
&= \int (\hat{\phi} * \hat{\mu})\psi dx.
\end{aligned}$$

The last line may be seen by writing out the definition of the convolution and using Fubini's theorem. Since this worked for all $\psi \in \mathcal{S}$, we get (74).

LEMMA 7.2. *Let* $f, g \in \mathcal{S}$, *and let* μ *be a (say) compactly supported measure. Then*

$$(75) \qquad \int f\bar{g} d\mu = \int (\hat{\mu} * \bar{g}) \cdot f dx.$$

PROOF. Recall that

$$\hat{\bar{g}} = \overline{\check{g}},$$

so that

$$\hat{\bar{\check{g}}} = \hat{\check{g}} = \bar{g}$$

by the inversion theorem. Now apply the duality relation and (74), obtaining

$$\int \widetilde{f}\overline{\hat{g}}d\mu \;=\; \int f\cdot(\overline{\hat{g}}\mu)\hat{}dx$$

$$=\; \int f\cdot(\overline{g}*\hat{\mu})dx$$

as claimed. $\qquad\square$

LEMMA 7.3. *Let μ be a finite positive measure. The following are equivalent for any q and any C.*

   (1) $\|\widehat{fd\mu}\|_q \le C\|f\|_2$, $f \in L^2(d\mu)$.
   (2) $\|\hat{g}\|_{L^2(d\mu)} \le C\|g\|_{q'}$, $g \in \mathcal{S}$.
   (3) $\|\hat{\mu}*f\|_q \le C^2\|f\|_{q'}$, $f \in \mathcal{S}$.

PROOF. Let $g \in \mathcal{S}$, $f \in L^2(d\mu)$. By the duality relation

$$(76) \qquad \int \hat{g}f d\mu = \int \widehat{fd\mu}\cdot g dx.$$

If 1. holds, then the right side of (76) is $\le \|g\|_{q'}\|\widehat{fd\mu}\|_q \le C\|g\|_{q'}\|f\|_{L^2(d\mu)}$ for any $f \in L^2(d\mu)$. Hence so is the left side. This proves 2. by duality. If 2. holds then the left side is $\le \|\hat{g}\|_{L^2(d\mu)}\|f\|_{L^2(d\mu)} \le C\|g\|_{q'}\|f\|_{L^2(d\mu)}$ for $g \in \mathcal{S}$. Hence so is the right side. Since $\mathcal{S}$ is dense in $L^{q'}$, this proves 1. by duality.

If 3. holds, then the right side of (75) is $\le C^2\|f\|_{q'}^2$ when $f = g \in \mathcal{S}$. Hence so is the left side, which proves 2. If 2. holds then, for any $f, g \in \mathcal{S}$, using also the Schwartz inequality the left side of (75) is $\le C^2\|f\|_{q'}\|g\|_{q'}$. Hence the right side of (75) is also $\le C^2\|f\|_{q'}\|g\|_{q'}$, which proves 3. by duality. $\qquad\square$

REMARK One can fit lemma 7.3 into the abstract setup

$$T : L^2 \to L^q \Leftrightarrow T^* : L^{q'} \to L^2 \Leftrightarrow TT^* : L^{q'} \to L^q.$$

This is the standard way to think about the lemma, although it is technically a bit easier to present the proof in the above ad hoc manner. Namely, if T is the operator $f \to \widehat{fd\sigma}$ then T^* is the operator $f \to \hat{f}$, where we regard $\hat{f}$ as being defined on the measure space associated to μ, and TT^* is convolution with $\hat{\mu}$.

PROOF OF THEOREM 7.1. We will not give a complete proof; we only prove (70) when $q > \frac{2n+2}{n-1}$ instead of $\ge$. For the endpoint, see for example [**35**], [**9**], [**32**], [**33**].

We will show that if $q > \frac{2n+2}{n-1}$, then

$$(77) \qquad \|\hat{\sigma}*f\|_q \le C_q\|f\|_{q'},$$

The relevant properties of σ will be

$$(78) \qquad \sigma(D(x,r)) \lesssim r^{n-1},$$

which reflects the $n-1$-dimensionality of the sphere, and the bound

$$(79) \qquad |\hat{\sigma}(\xi)| \lesssim (1+|\xi|)^{-\frac{n-1}{2}}$$

from Corollary 6.6.

Let ϕ be a C^∞ function with the following properties:

$$\operatorname{supp}\phi \subset \{x : \frac{1}{4} \le x \le 1\},$$

$$\text{if } |x| \ge 1 \text{ then } \sum_{j\ge 0}\phi(2^{-j}x) = 1.$$

Such a function may be obtained as follows: let χ be a C^∞ function which is equal to 1 when $|x| \ge 1$ and to 0 when $|x| \le \frac{1}{2}$, and let $\phi(x) = \chi(2x) - \chi(x)$.

We now cut up $\hat{\sigma}$ as follows:

$$\hat{\sigma} = K_{-\infty} + \sum_{j=0}^{\infty} K_j,$$

where

$$K_j(x) = \phi(2^{-j}x)\hat{\sigma}(x),$$

$$K_{-\infty}(x) = (1 - \sum_{j=0}^{\infty}\phi(2^{-j}x))\hat{\sigma}.$$

Then $K_{-\infty}$ is a C_0^∞ function, so

$$\|K_{-\infty} * f\|_q \lesssim \|f\|_p$$

by Young's inequality, provided $q \ge p$. In particular, since $q > 2$ we may take $p = q'$.

We now consider the terms in the sum. The logic will be that we estimate convolution with K_j as an operator from L^1 to L^∞ and from L^2 to L^2, and then use Riesz-Thorin. We have

$$\|K_j\|_\infty \lesssim 2^{-j\frac{n-1}{2}}$$

by (79). Using the trivial bound $\|K_j * f\|_\infty \le \|K_j\|_\infty\|f\|_1$ we conclude our $L^1 \to L^\infty$ bound,

$$(80) \qquad \|K_j * f\|_\infty \lesssim 2^{-j\frac{n-1}{2}}\|f\|_1.$$

On the other hand, we can use (78) to estimate $\widehat{K_j}$. Namely, let $\psi = \hat{\phi}$. Note also that $\hat{\sigma} = \check{\sigma}$, since σ and therefore $\hat{\sigma}$ are invariant under the reflection $x \to -x$. Accordingly we have

$$\widehat{K_j} = \psi^{2^{-j}} * \sigma,$$

using (73) and the fact that $\widehat{\hat{\phi}_\epsilon} = \hat{\phi}^\epsilon$. Since $\psi \in \mathcal{S}$, it follows that

$$|\widehat{K_j}(\xi)| \le C_N 2^{jn}\int(1+2^j|\xi-\eta|)^{-N}d\sigma(\eta)$$

for any fixed $N < \infty$. Therefore

$$
\begin{aligned}
|\widehat{K_j}(\xi)| \;\leq\;& C_N 2^{jn} \left(\int_{D(\xi,2^{-j})} (1 + 2^j|\xi - \eta|)^{-N} d\sigma(\eta) \right. \\
&\left. + \sum_{k\geq 0} \int_{D(\xi,2^{k+1-j})\setminus D(\xi,2^{k-j})} (1 + 2^j|\xi - \eta|)^{-N} d\sigma(\eta) \right) \\
\leq\;& C_N 2^{jn} \left(\sigma(D(\xi,2^{-j})) + \sum_{k\geq 0} 2^{-Nk}\sigma(D(\xi,2^{k+1-j})\setminus D(\xi,2^{k-j})) \right) \\
\lesssim\;& 2^{jn} \left(2^{-j(n-1)} + \sum_{k\geq 0} 2^{-Nk} 2^{(n-1)(k-j)} \right) \\
\lesssim\;& 2^j,
\end{aligned}
$$

where we used (78) at the next to last line, and at the last line we fixed N to be equal to n and summed a geometric series. Thus

$$
(81) \qquad\qquad \|\widehat{K_j}\|_\infty \lesssim 2^j.
$$

Now we mention the trivial but important fact that

$$
(82) \qquad\qquad \|K * f\|_2 \leq \|\hat{K}\|_\infty \|f\|_2
$$

if, say, K and f are in $\mathcal{S}$. This follows since

$$
\begin{aligned}
\|K * f\|_2 \;&=\; \|\widehat{K * f}\|_2 \\
&=\; \|\hat{K}\hat{f}\|_2 \\
&\leq\; \|\hat{K}\|_\infty \|\hat{f}\|_2 \\
&=\; \|\hat{K}\|_\infty \|f\|_2.
\end{aligned}
$$

Combining (81) and (82) we conclude that

$$
(83) \qquad\qquad \|K_j * f\|_2 \lesssim 2^j \|f\|_2.
$$

Accordingly, by (80), (83) and Riesz-Thorin we have

$$
\|K_j * f\|_q \lesssim 2^{j\theta} 2^{-j\frac{n-1}{2}(1-\theta)} \|f\|_{q'}
$$

if $\frac{\theta}{2} + \frac{1-\theta}{\infty} = \frac{1}{q}$. This works out to

$$
(84) \qquad\qquad \|K_j * f\|_q \lesssim 2^{j(\frac{n+1}{q} - \frac{n-1}{2})} \|f\|_{q'}
$$

for any $q \in [2, \infty]$. If $q > \frac{2n+2}{n-1}$ then the exponent $\frac{n+1}{q} - \frac{n-1}{2}$ is negative, so we conclude that

$$
\sum_j \|K_j * f\|_q \lesssim \|f\|_{q'}.
$$

Since f is a Schwartz function, the sum

$$K_{-\infty} * f + \sum_j K_j * f$$

is easily seen to converge pointwise to $\hat{\sigma} * f$. We conclude using Fatou's lemma that $\|\hat{\sigma} * f\|_q \lesssim \|f\|_{q'}$, as claimed. $\qquad\square$

FURTHER REMARKS 1. Notice that the L^2 estimate in the preceding argument was based only on dimensionality considerations. This suggests that there should be an L^2 bound for $\widehat{f d\sigma}$ valid under very general conditions.

THEOREM 7.4. *Let ν be a positive finite measure satisfying the estimate*

$$(85) \qquad \nu(D(x,r)) \leq C r^\alpha.$$

Then there is a bound

$$(86) \qquad \|\widehat{f d\nu}\|_{L^2(D(0,R))} \leq C R^{\frac{n-\alpha}{2}} \|f\|_{L^2(d\nu)}.$$

The proof uses the following "generic" test for L^2 boundedness.

LEMMA 7.5 (Schur's test). *Let (X, μ) and (Y, ν) be measure spaces, and let $K(x, y)$ be a measurable function on $X \times Y$ with*

$$(87) \qquad \int_X |K(x,y)| d\mu(x) \leq A \text{ for each } y,$$

$$(88) \qquad \int_Y |K(x,y)| d\nu(y) \leq B \text{ for each } x.$$

Define $T_K f(x) = \int K(x,y) f(y) d\nu(y)$. Then for $f \in L^2(d\nu)$ the integral defining $T_K f$ converges a.e. $(d\mu(x))$ and there is an estimate

$$(89) \qquad \|T_K f\|_{L^2(d\mu)} \leq \sqrt{AB} \|f\|_{L^2(d\nu)}.$$

PROOF. It is possible to use Riesz-Thorin here, since (88) implies $\|T_K f\|_\infty \leq B\|f\|_\infty$ and (87) implies $\|T_K f\|_1 \leq A\|f\|_1$.

A more "elementary" argument goes as follows. If a and b are positive numbers then we have

$$(90) \qquad \sqrt{ab} = \min_{\epsilon \in (0,\infty)} \frac{1}{2}(\epsilon a + \epsilon^{-1} b),$$

since $\leq$ is the arithmetic-geometric mean inequality and $\geq$ follows by taking $\epsilon = \sqrt{\frac{b}{a}}$.

To prove (89) it suffices to show that if $\|f\|_{L^2(d\mu)} \leq 1$, $\|g\|_{L^2(d\nu)} \leq 1$, then

$$(91) \qquad \int\int |K(x,y)||f(x)||g(y)| d\mu(x) d\nu(y) \leq \sqrt{AB}.$$

To show (91), we estimate

$$\int\int \ |K(x,y)||f(x)||g(y)|d\mu(x)d\nu(y)$$

$$= \frac{1}{2}\min_{\epsilon}\left(\epsilon\int\int |K(x,y)||g(y)|^2 d\mu(x)d\nu(y)\right.$$

$$\left. +\epsilon^{-1}\int\int |K(x,y)||f(x)|^2 d\nu(y)d\mu(x)\right)$$

$$\leq \frac{1}{2}\min_{\epsilon}\left(\epsilon A\int |g(y)|^2 d\nu(y) + \epsilon^{-1}B\int |f(x)|^2 d\mu(x)\right)$$

$$\leq \frac{1}{2}\min_{\epsilon}(\epsilon A + \epsilon^{-1}B)$$

$$= \sqrt{AB}.$$

$\square$

To prove Theorem 7.4, let ϕ be an even Schwartz function which is ≥ 1 on the unit disc and whose Fourier transform has compact support. (Exercise: show that such a function exists.) In the usual way define $\phi_{R^{-1}}(x) = \phi(R^{-1}x)$. Then

$$\|\widehat{fd\nu}\|_{L^2(D(0,R))} \ \leq \ \|\phi_{R^{-1}}(x)\widehat{fd\nu}(-x)\|_{L^2(dx)}$$

$$= \ \|\widehat{\phi_{R^{-1}}} * (fd\nu)\|_2$$

by (73) and Plancherel.

The last line is the $L^2(dx)$ norm of the function

$$\int R^n\hat{\phi}(R(x-y))f(y)d\nu(y).$$

We have the estimates

$$\int |R^n\hat{\phi}(R(x-y))|dx = \|\hat{\phi}\|_1 < \infty$$

for each fixed y, by change of variables, and

$$\int |R^n\hat{\phi}(R(x-y))|d\nu(y) \lesssim R^{n-\alpha}$$

for each fixed x, by (85) and the compact support of $\hat{\phi}$. By Lemma 7.5

$$\|\int R^n\hat{\phi}(R(x-y))f(y)d\nu(y)\|_{L^2(dx)} \lesssim R^{\frac{n-\alpha}{2}}\|f\|_{L^2(d\nu)},$$

and the proof is complete. $\square$

2. Another remark is that it is possible to base the whole proof of Theorem 7.1 on the stationary phase asymptotics in section 6, instead of explicitly using the dimensionality of σ. This sort of argument has the obvious advantage that it is more flexible, since it works also in other situations where the "convolution kernel" $\hat{\sigma}(x-y)$ is replaced by a kernel $K(x,y)$ satisfing appropriate conditions. See for example [**29**], [**32**], [**33**]. On the other hand,

it is more complicated and is not as relevant in connection with more delicate questions such as the restriction conjecture, which is known to be false in most of the more general situations (see [5], [26], [30]). We give a brief sketch omitting details. The basic result is the so-called variable coefficient Plancherel theorem, due to Hörmander [16].

Let ϕ be a real valued C^∞ function defined on $\mathbb{R}^n \times \mathbb{R}^n$, let $a \in C_0^\infty(\mathbb{R}^n \times \mathbb{R}^n)$ and consider the "oscillatory integral operators"

$$(92) \qquad T_\lambda f(x) = \int e^{-\pi i \lambda \phi(x,y)} a(x,y) f(y) dy.$$

Since a has compact support, it is obvious that these map $L^2(\mathbb{R}^n)$ to $L^2(\mathbb{R}^n)$ with a norm bound independent of λ, but we want to show that the norm decays in a suitable way as $\lambda \to \infty$. As with the oscillatory integrals of section 6, this will not be the case if ϕ is too degenerate. In the present situation, note that if ϕ depends on x only, then the factor $e^{-\pi i \lambda \phi}$ in (92) may be taken outside the integral sign, so the norm is independent of λ. Similarly, if ϕ depends on y only, then the factor $e^{-\pi i \lambda \phi}$ may be incorporated into f. We conclude in fact that if $\phi(x,y) = a(x) + b(y)$, then $\|T_\lambda\|_{L^2 \to L^2}$ is independent of λ. This strongly suggests that the appropriate nondegeneracy condition should involve the "mixed Hessian"

$$\tilde{H}_\phi = \left(\frac{\partial^2 \phi}{\partial x_i \partial y_j} \right)_{i,j=1}^n$$

since the mixed Hessian vanishes identically if $\phi(x,y) = a(x) + b(y)$.

THEOREM A (Hörmander) *Assume that*

$$det\left(\tilde{H}_\phi(x,y) \right) \neq 0$$

at all points $(x,y) \in \operatorname{supp} a$. *Then*

$$\|T_\lambda\|_{L^2 \to L^2} \leq C \lambda^{-\frac{n}{2}}.$$

SKETCH OF PROOF This is evidently related to stationary phase, but one cannot apply stationary phase directly to the integral (92), since f isn't smooth. Instead, one looks at $T_\lambda T_\lambda^*$ which is an integral operator T_K with kernel

$$(93) \qquad K(x,y) = \int e^{-\pi i \lambda(\phi(x,z) - \phi(y,z))} a(x,z)\overline{a(y,z)} dz.$$

The assumption about the mixed Hessian guarantees that the phase function in (93) has no critical points if x and y are close together. Using a version[1] of "nonstationary phase" one can obtain the estimate

$$\forall N \; \exists C_N : |K(x,y)| \leq C_N (1 + \lambda |x - y|)^{-N},$$

---

[1] One needs something a bit more quantitative than our Proposition 6.1; the necessary lemma is best proved by integration by parts. See for example [32].

provided $|x - y|$ is less than a suitable constant. It follows that if a has small support then

$$\int |K(x,y)|\,dy \lesssim \lambda^{-n}$$

for each fixed x, and similarly

$$\int |K(x,y)|\,dx \lesssim \lambda^{-n}$$

for each y. Then Schur's test shows that $\|T_\lambda T_\lambda^*\|_{L^2 \to L^2} \lesssim \lambda^{-n}$, so $\|T_\lambda\|_{L^2 \to L^2} \lesssim \lambda^{-\frac{n}{2}}$. The small support assumption on a can then be removed using a partition of unity. $\square$.

It is possible to generalize this to the case where the rank of $\tilde{H}_\phi$ is $\geq k$, where $k \in \{1, \ldots, n\}$; just replace the exponent $\frac{n}{2}$ by $\frac{k}{2}$. Furthermore, the compact support assumption on a may be replaced by "proper support" (see the statement below), and finally one can obtain $L^{q'} \to L^q$ estimates by interpolating with the trivial $\|T_\lambda\|_{L^1 \to L^\infty} \leq 1$. Here then is the variable coefficient Plancherel, souped up in a manner which makes it applicable in connection with Stein-Tomas. See the references mentioned above.

THEOREM B (Hörmander) *Assume that a is a C^∞ function supported on the set $\{(x,y) \in \mathbb{R}^n \times \mathbb{R}^n : |x - y| \leq C\}$ whose all partial derivatives are bounded. Let ϕ be a real valued C^∞ function defined on a neighborhood of $\operatorname{supp} a$, all of whose partial derivatives are bounded, and such that the rank of $\tilde{H}_\phi(x,y)$ is at least k everywhere. Assume furthermore that the sum of the absolute values of the determinants of the k by k minors of $\tilde{H}_\phi$ is bounded away from zero. Then there is a bound*

$$\|T_\lambda\|_{L^{q'} \to L^q} \leq C\lambda^{-\frac{k}{q}}$$

when $2 \leq q \leq \infty$.

Now look back at the proof we gave for Theorem 7.1. The main point was to obtain the bound (84). Now, $K_j(x)$ is the real part of

$$\tilde{K}_j(x) \overset{def}{=} \phi(2^{-j}x)a(|x|)e^{-2\pi i|x|},$$

where a satisfies the estimates in Corollary 6.6. Accordingly, it suffices to prove (84) with K_j replaced by $\tilde{K}_j$. Let T_j be convolution with $\tilde{K}_j$, and rescale by 2^j; thus we consider the operator

$$f \to T_j(f_{2^j})_{2^{-j}}.$$

This is an integral operator S_j whose kernel is

$$2^{nj}\phi(x - y)a(2^j|x - y|)e^{-2\pi i 2^j|x-y|}.$$

We want to apply Theorem B to S_j; toward this end we make the following observations.

(i) From the estimates in Corollary 6.6, we see that the functions

$$2^{\frac{n-1}{2}j}\phi(x - y)a(2^j|x - y|)$$

have derivative bounds which are independent of j, and clearly they are supported in $\frac{1}{4} \le |x - y| \le 1$.

(ii) The mixed Hessian of the function $|x - y|$ has rank $n - 1$. This is a calculation which we leave to the reader, just noting that the exceptional direction corresponds to the direction along the line segment $\overline{xy}$.

It follows that the operators $2^{-\frac{n+1}{2}j}S_j$ satisfy the hypotheses of Theorem B with $k = n - 1$, uniformly in j, if we take $\lambda = 2^{j+1}$. Accordingly,

$$\|S_j f\|_q \lesssim 2^{(\frac{n+1}{2} - \frac{n-1}{q})j}\|f\|_{q'},$$

and therefore using change of variables

$$2^{-\frac{n}{q}j}\|T_j f\|_q \lesssim 2^{(\frac{n+1}{2} - \frac{n-1}{q})j}2^{-\frac{n}{q'}j}\|f\|_{q'},$$

i.e.

$$\|T_j f\|_q \lesssim 2^{(\frac{n+1}{q} - \frac{n-1}{2})j}\|f\|_{q'},$$

which is (84). $\qquad\qquad\qquad\qquad\qquad\qquad\qquad\qquad\qquad\qquad\qquad\square$

EXERCISE: Use Theorem A for an appropriate phase function, and a rescaling argument of the preceding type, to prove the bound

$$\|\hat{f}\|_2 \le C\|f\|_2.$$

This explains the name "variable coefficient Plancherel theorem".

CHAPTER 8

Hausdorff Measures

Fix $\alpha > 0$, and let $E \subset \mathbb{R}^n$. For $\epsilon > 0$, one defines

$$H_\alpha^\epsilon(E) = \inf(\sum_{j=1}^\infty r_j^\alpha),$$

where the infimum is taken over all countable coverings of E by discs $D(x_j, r_j)$ with $r_j < \epsilon$. It is clear that $H_\alpha^\epsilon(E)$ increases as ϵ decreases, and we define

$$H_\alpha(E) = \lim_{\epsilon \to 0} H_\alpha^\epsilon(E).$$

It is also clear that $H_\alpha^\epsilon(E) \leq H_\beta^\epsilon(E)$ if $\alpha > \beta$ and $\epsilon \leq 1$; thus

$$(94) \qquad H_\alpha(E) \text{ is a nonincreasing function of } \alpha.$$

REMARKS 1. If $H_\alpha^1(E) = 0$, then $H_\alpha(E) = 0$. This follows readily from the definition, since a covering showing that $H_\alpha^1(E) < \delta$ will necessarily consist of discs of radius $< \delta^{\frac{1}{\alpha}}$.

2. It is also clear that $H_\alpha(E) = 0$ for all E if $\alpha > n$, since one can then cover $\mathbb{R}^n$ by discs $D(x_j, r_j)$ with $\sum_j r_j^\alpha$ arbitrarily small.

LEMMA 8.1. *There is a unique number α_0, called the Hausdorff dimension of E or $\dim E$, such that $H_\alpha(E) = \infty$ if $\alpha < \alpha_0$ and $H_\alpha(E) = 0$ if $\alpha > \alpha_0$.*

PROOF. Define α_0 to be the supremum of all α such that $H_\alpha(E) = \infty$. Thus $H_\alpha(E) = \infty$ if $\alpha < \alpha_0$, by (94). Suppose $\alpha > \alpha_0$. Let $\beta \in (\alpha_0, \alpha)$. Define $M = 1 + H_\beta(E) < \infty$. If $\epsilon > 0$, then we have a covering by discs with $\sum_j r_j^\beta \leq M$ and $r_j < \epsilon$. So

$$\sum_j r_j^\alpha \leq \epsilon^{\alpha - \beta} \sum_j r_j^\beta \leq \epsilon^{\alpha - \beta} M$$

which goes to 0 as $\epsilon \to 0$. Thus $H_\alpha(E) = 0$. $\qquad \square$.

FURTHER REMARKS 1. The set function H_α may be seen to be countably additive on Borel sets, i.e. defines a Borel measure. See standard references in the area like [6], [10], [25]. This is part of the reason one considers H_α instead of, say, H_α^1. Notice in this connection that if E and F are disjoint compact sets, then evidently $H_\alpha(E \cup F) = H_\alpha(E) + H_\alpha(F)$. This statement is already false for H_α^1.

57

2. The Borel measure H_n coincides with $\frac{1}{\omega}$ times Lebesgue measure, where ω is the volume of the unit ball. If $\alpha < n$, then H_α is non-sigma finite; this follows e.g. by Lemma 8.1, which implies that any set with nonzero Lebesgue measure will have infinite H_α-measure.

EXAMPLES The canonical example is the usual $\frac{1}{3}$-Cantor set on the line. This has a covering by 2^n intervals of length 3^{-n}, so it has finite $H_{\frac{\log 2}{\log 3}}$-measure. It is not difficult to show that in fact its $H_{\frac{\log 2}{\log 3}}$-measure is nonzero; this can be done geometrically, or one can apply Proposition 8.2 below to the Cantor measure. In particular, the dimension of the Cantor set is $\frac{\log 2}{\log 3}$.

Now consider instead a Cantor set with variable dissection ratios $\{\epsilon_n\}$, i.e. one starts with the interval $[0,1]$, removes the middle ϵ_1 proportion, then removes the middle ϵ_2 proportion of each of the resulting intervals and so forth. If we assume that $\epsilon_{n+1} \le \epsilon_n$, and let $\epsilon = \lim_{n\to\infty} \epsilon_n$, then it is not hard to show that the dimension of the resulting set E will be $\frac{\log 2}{\log(\frac{2}{1-\epsilon})}$. In particular, if $\epsilon_n \to 0$ then $\dim E = 1$. On the other hand, $H_1(E)$ will be positive only if $\sum_n \epsilon_n < \infty$, so this gives examples of sets with zero Lebesgue measure but "full" Hausdorff dimension.

There are numerous other notions of dimension. We mention only one of them, the Minkowski dimension, which we define here only for compact sets. Namely, if E is compact then let $E_\delta = \{x \in \mathbb{R}^n : \mathrm{dist}(x,E) < \delta\}$. Let α_0 be the supremum of all numbers α such that, for some constant C,

$$|E_\delta| \ge C\delta^{n-\alpha}$$

for all $\delta \in (0,1]$. Then α_0 is called the lower Minkowski dimension and denoted $d_L(E)$. Let α_1 be the supremum of all numbers α such that, for some constant C,

$$|E_\delta| \ge C\delta^{n-\alpha}$$

for a sequence of δ's which converges to zero. Then α_1 is called the upper Minkowski dimension and denoted $d_U(E)$.

It would also be possible to define these like Hausdorff dimension but restricting to coverings by discs all the same size, namely: define a set S to be δ-separated if any two distict points $x, y \in S$ satisfy $|x - y| > \delta$. Let $\mathcal{E}_\delta(E)$ ("δ-entropy on E") be the maximal possible cardinality for a δ-separated subset[1] of E. Then it is not hard to show that

$$d_L(E) = \liminf_{\delta\to 0} \frac{\log \mathcal{E}_\delta(E)}{\log \frac{1}{\delta}},$$

$$d_U(E) = \limsup_{\delta\to 0} \frac{\log \mathcal{E}_\delta(E)}{\log \frac{1}{\delta}}.$$

---

[1]Exercise: show that $\mathcal{E}_\delta(E)$ is comparable to the minimum number of δ-discs required to cover E

Notice that a countable set may have positive lower Minkowski dimension; for example, the set $\{\frac{1}{n}\}_{n=1}^{\infty} \cup \{0\}$ has upper and lower Minkowski dimension $\frac{1}{2}$.

If E is a compact set, then let $P(E)$ be the space of the probability measures supported on E.

PROPOSITION 8.2. *Suppose $E \subset \mathbb{R}^n$ is compact. Assume that there is a $\mu \in P(E)$ with*

$$(95) \qquad\qquad \mu(D(x,r)) \leq C r^\alpha$$

for a suitable constant C and all $x \in \mathbb{R}^n$, $r > 0$. Then $H_\alpha(E) > 0$. Conversely, if $H_\alpha(E) > 0$, there is a $\mu \in P(E)$ such that (95) holds.

PROOF. The first part is easy: let $\{D(x_j, r_j)\}$ be any covering of E by discs. Then

$$1 = \mu(E) \leq \sum_j \mu(D(x_j, r_j)) \leq C \sum_j r_j^\alpha,$$

which shows that $H_\alpha(E) \geq C^{-1}$.

The proof of the converse involves constructing a suitable measure, which is most easily done using dyadic cubes. Thus we let $\mathcal{Q}_k$ be all cubes of side length $\ell(Q) = 2^{-k}$ whose vertices are at points of $2^{-k}\mathbb{Z}^n$. We can take these to be closed cubes, for definiteness. It is standard to work with these in such contexts because of their nice combinatorics: if $Q \in \mathcal{Q}_k$, then there is a unique $\tilde{Q} \in \mathcal{Q}_{k-1}$ with $Q \subset \tilde{Q}$; furthermore if we fix $Q_1 \in \mathcal{Q}_{k-1}$, then Q_1 is the union of those $Q \in \mathcal{Q}_k$ with $\tilde{Q} = Q_1$, and the union is disjoint except for edges. A *dyadic cube* is a cube which is in $\mathcal{Q}_k$ for some k.

If Q is a dyadic cube, then clearly there is a disc $D(x,r)$ with $Q \subset D(x,r)$ and $r \leq C\ell(Q)$. Likewise, if we fix $D(x,r)$, then there are a bounded number of dyadic cubes $Q_1 \dots Q_C$ with $\ell(Q_j) \leq Cr$ and whose union contains $D(x,r)$. From these properties, it is easy to see that the definition of Hausdorff measure and also the property (95) could equally well be given in terms of dyadic cubes. Thus, except for the values of the constants,

$$\mu \text{ satisfies } (95) \Leftrightarrow \mu(Q) \leq C\ell(Q)^\alpha \text{ for all dyadic cubes } Q.$$

Furthermore, if we define

$$h_\alpha^\epsilon(E) = \inf\Big(\sum_{Q \in \mathcal{F}} \ell(Q)^\alpha : E \subset \bigcup_{Q \in \mathcal{F}} Q \Big),$$

where $\mathcal{F}$ runs over all coverings of E by dyadic cubes of side length $\ell(Q) < \epsilon$, and

$$h_\alpha(E) = \lim_{\epsilon \to 0} h_\alpha^\epsilon(E),$$

then we have

$$C^{-1} H_\alpha^\epsilon(E) \leq h_\alpha^\epsilon(E) \leq C H_\alpha^\epsilon(E),$$

therefore

$$h_\alpha(E) > 0 \Leftrightarrow H_\alpha(E) > 0.$$

We return now to the proof of Proposition 8.2. We may assume that E is contained in the unit cube $[0,1] \times \ldots \times [0,1]$. By the preceding remarks and Remark 1. above we may assume that $h_\alpha^1(E) > 0$, and it suffices to find $\mu \in P(E)$ so that $\mu(Q) \le C\ell(Q)^\alpha$ for all dyadic cubes Q with $\ell(Q) \le 1$. We now make a further reduction.

CLAIM. It suffices to find, for each fixed $m \in \mathbb{Z}^+$, a positive measure μ with the following properties:

(96) μ is supported on the union of the cubes $Q \in \mathcal{Q}_m$ which intersect E;

$$\text{(97)} \qquad\qquad\qquad\qquad \|\mu\| \ge C^{-1};$$

$$\text{(98)} \qquad \mu(Q) \le \ell(Q)^\alpha \text{ for all dyadic cubes with } \ell(Q) \ge 2^{-m}.$$

Here C is independent of m.

Namely, if this can be done, then denote the measures satisfying (96), (97), (98) by μ_m. (98) implies a bound on $\|\mu_m\|$, so there is a weak* limit point μ. (96) then shows that μ is supported on E, (98) shows that $\mu(Q) \le \ell(Q)^\alpha$ for all dyadic cubes, and (97) shows that $\|\mu\| \ge C^{-1}$. Accordingly, a suitable scalar multiple of μ gives us the necessary probability measure.

There are a number of ways of constructing the measures satisfying (96), (97), (98). Roughly, the issue is that (97) and (98) are competing conditions, and one has to find a measure μ with the appropriate support and with total mass roughly as large as possible given that (97) holds. This can be done for example by using finite dimensional convexity theory (exercise!). We present a different (more constructive) argument taken from [6], Chapter 2.

We fix m, and will construct a finite sequence of measures $\nu_m, \ldots, \nu_0$, in that order; ν_0 will be the measure we want.

Start by defining ν_m to be the unique measure with the following properties.

1. On each $Q \in \mathcal{Q}_m$, ν_m is a scalar multiple of Lebesgue measure.
2. If $Q \in \mathcal{Q}_m$ and $Q \cap E = \emptyset$, then $\nu_m(Q) = 0$.
3. If $Q \in \mathcal{Q}_m$ and $Q \cap E \ne \emptyset$, then $\nu_m(Q) = 2^{-m\alpha}$.

If we set $k = m$, then ν_k has the following properties: it is absolutely continuous with respect to the Lebesgue measure, and

(A) $\nu_k(Q) \le \ell(Q)^\alpha$ if Q is a dyadic cube with side 2^{-j}, $k \le j \le m$;

(B) if Q_1 is a dyadic cube of side 2^{-k}, then there is a covering $\mathcal{F}_{Q_1}$ of $Q_1 \cap E$ by dyadic cubes contained in Q_1 such that $\nu_k(Q_1) \ge \sum_{Q \in \mathcal{F}_{Q_1}} \ell(Q)^\alpha$.

Assume now that $1 \le k \le m$ and we have constructed an absolutely continuous measure ν_k with properties (96), (A) and (B). We will construct ν_{k-1} having these same properties, where in (A) and (B) k is replaced by $k-1$. Namely, to define ν_{k-1} it suffices to define $\nu_{k-1}(Y)$ when Y is contained in a cube $Q \in \mathcal{Q}_{k-1}$. Fix $Q \in \mathcal{Q}_{k-1}$. Consider two cases.

(i) $\nu_k(Q) \le \ell(Q)^\alpha$. In this case we let ν_{k-1} agree with ν_k on subsets of Q.

(ii) $\nu_k(Q) \geq \ell(Q)^\alpha$. In this case we let ν_{k-1} agree with $c\nu_k$ on subsets of Q, where c is the scalar $\frac{2^{-(k-1)\alpha}}{\nu_k(Q)}$.

Notice that $\nu_{k-1}(Y) \leq \nu_k(Y)$ for any set Y, and furthermore $\nu_{k-1}(Q) \leq \ell(Q)^\alpha$ if $Q \in \mathcal{Q}_{k-1}$. These properties and (A) for ν_k give (A) for ν_{k-1}, and (96) for ν_{k-1} follows trivially from (96) for ν_k. To see (B) for ν_{k-1}, fix $Q \in \mathcal{Q}_{k-1}$. If Q is as in case (ii), then $\nu_{k-1}(Q) = \ell(Q)^\alpha$, so we can use the covering by the singleton $\{Q\}$. If Q is as in case (i), then for each of the cubes $Q_j \in \mathcal{Q}_k$ whose union is Q we have the covering of $Q_j \cap E$ associated with (100) for ν_k. Since ν_k and ν_{k-1} agree on subsets of Q, we can simply put these coverings together to obtain a suitable covering of $Q \cap E$. This concludes the inductive step from ν_k to ν_{k-1}.

We therefore have constructed ν_0. It has properties (96), (98) (since for ν_0 this is equivalent to (A)), and by (B) and the definition of h_α^1 it has property (97). $\qquad\square$

Let us now define the *α-dimensional energy* of a (positive) measure μ with compact support[2] by the formula

$$I_\alpha(\mu) = \int\int |x - y|^{-\alpha} d\mu(x)d\mu(y).$$

We always assume that $0 < \alpha < n$. We also define

$$V_\mu^\alpha(y) = \int |x - y|^{-\alpha} d\mu(x).$$

Thus

$$(99) \qquad I_\alpha(\mu) = \int V_\mu^\alpha d\mu.$$

The "potential" V_μ^α is very important in other contexts (namely elliptic theory, since it is harmonic away from $\operatorname{supp}\mu$ when $\alpha = n - 2$) but less important than the energy here. Nevertheless we will use it in a technical way below. Notice that it is actually the convolution of the function $|x|^{-\alpha}$ with the measure μ.

Roughly, one expects a measure to have $I_\alpha(\mu) < \infty$ if and only if it satisfies (95); this precise statement is false, but we see below that nevertheless the Hausdorff dimension of a compact set can be defined in terms of the energies of measures in $P(E)$.

LEMMA 8.3. *(i) If μ is a probability measure with compact support satisfying (95), then $I_\beta(\mu) < \infty$ for all $\beta < \alpha$.*

(ii) Conversely, if μ is a probability measure with compact support and with $I_\alpha(\mu) < \infty$, then there is another probability measure ν such that $\nu(X) \leq 2\mu(X)$ for all sets X and such that ν satisfies (95).

---

[2]The compact support assumption is not needed; it is included to simplify the presentation.

PROOF. (i) We can assume that the diameter of the support of μ is ≤ 1. Then

$$V_\mu^\beta(x) \lesssim \sum_{j=0}^\infty 2^{j\beta}\mu(D(x,2^{-j})).$$

Accordingly, if μ satisfies (95), and $\beta < \alpha$, then

$$V_\mu^\beta(x) \lesssim \sum_{j=0}^\infty 2^{j\beta}2^{-j\alpha}$$
$$\lesssim 1.$$

It follows by (99) that $I_\alpha(\mu) < \infty$.

(ii) Let F be the set of points x such that $V_\mu^\alpha(x) \leq 2I_\alpha(\mu)$. Then $\mu(F) \geq \frac{1}{2}$ by (99). Let χ_F be the indicator function of F and let $\nu(X) = \mu(X \cap F)/\mu(F)$. We need to show that ν satisfies (95). Suppose first that $x \in F$. If $r > 0$ then

$$r^{-\alpha}\nu(D(x,r)) \leq V_\nu^\alpha(x) \leq 2V_\mu^\alpha(x) \leq 4I_\alpha(\mu).$$

This verifies (95) when $x \in F$. For general x, consider two cases. If r is such that $D(x,r) \cap F = \emptyset$ then evidently $\nu(D(x,r)) = 0$. If $D(x,r) \cap F \neq \emptyset$, let $y \in D(x,r) \cap F$. Then $\nu(D(x,r)) \leq \nu(D(y,2r)) \lesssim r^\alpha$ by the first part of the proof. $\qquad\square$

PROPOSITION 8.4. *If E is compact then the Hausdorff dimension of E coincides with the number*

$$\sup\{\alpha : \exists \mu \in P(E) \text{ with } I_\alpha(\mu) < \infty\}.$$

PROOF. Denote the above supremum by s. If $\beta < s$ then by (ii) of Lemma 8.3 E supports a measure with $\mu(D(x,r)) \leq Cr^\beta$. Then by Proposition 8.2 $H_\beta(E) > 0$, so $\beta \leq \dim E$. So $s \leq \dim E$. Conversely, if $\beta < \dim E$ then by Proposition 8.2 E supports a measure with $\mu(D(x,r)) \leq Cr^{\beta+\epsilon}$ for $\epsilon > 0$ small enough. Then $I_\beta(\mu) < \infty$, so $\beta \leq s$, which shows that $\dim E \leq s$. $\square$

The energy is a quadratic expression in μ and is therefore susceptible to Fourier transform arguments. Indeed, the following formula is essentially just Lemma 7.2 combined with the formula for the Fourier transform of $|x|^{-\alpha}$.

PROPOSITION 8.5. *Let μ be a positive measure with compact support and $0 < \alpha < n$. Then*

$$(100) \qquad \int\int |x-y|^{-\alpha}d\mu(x)d\mu(y) = c_\alpha \int |\hat{\mu}(\xi)|^2|\xi|^{-(n-\alpha)}d\xi,$$

where $c_\alpha = \dfrac{\gamma(\frac{n-a}{2})\pi^{a-\frac{n}{2}}}{\gamma(\frac{a}{2})}$.

PROOF. Suppose first that $f \in L^1$ is real and even, and that $d\mu(x) = \phi(x)dx$ with $\phi \in \mathcal{S}$. Then we have

$$
\text{(101)} \qquad \int f(x-y)d\mu(x)d\mu(y) = \int |\hat{\mu}(\xi)|^2 \hat{f}(\xi)d\xi
$$

This is proved like Lemma 7.2 using (73) instead of (74). Now fix ϕ. Then both sides of (101) are easily seen to define continuous linear maps from $f \in L^2$ to $\mathbb{R}$. Accordingly, (101) remains valid when $f \in L^1 + L^2, \phi \in \mathcal{S}$. Applying Proposition 4.1, we conclude (100) if $d\mu(x) = \phi(x)dx$, $\phi \in \mathcal{S}$. To pass to general measures, we use the following fact.

LEMMA 8.6. *Let ϕ be any radial decreasing Schwartz function with L^1 norm 1, and let $0 < \alpha < n$. Then*

$$
\int |x-y|^{-\alpha}\phi(y)dy \lesssim |x|^{-\alpha},
$$

where the implicit constant depends only on α, not on the choice of ϕ.

We sketch the proof as follows: one can easily reduce to the case where $\phi = \frac{1}{|D(0,R)|}\chi_{D(0,R)}$, and this case can be done by explicit calculation. $\qquad \square$

Now let $\phi(x) = e^{-\pi|x|^2}$. We have then $\phi^\epsilon * \mu \in \mathcal{S}$, so

$$
\text{(102)} \qquad
\begin{aligned}
&\int\int \left(\int\int |x-y|^{-\alpha}\phi^\epsilon(x-z)\phi^\epsilon(y-w)dxdy \right) d\mu(z)d\mu(w) \\
&= c_\alpha \int |\hat{\mu}(\xi)|^2 |\hat{\phi}(\epsilon\xi)|^2 |\xi|^{-(n-\alpha)}d\xi.
\end{aligned}
$$

Now let $\epsilon \to 0$. On the left side of (102), the expression inside the parentheses converges pointwise to $|z-w|^{-\alpha}$ using a minor variant on Lemma 3.2. If $I_\alpha(\mu) < \infty$ then the convergence is dominated in view of the preceding lemma, so the integrals on the left side converge to $I_\alpha(\mu)$. If $I_\alpha(\mu) = \infty$, then this remains true using Fatou's lemma. On the right hand side of (102) we can argue similarly: the integrands converge pointwise to $|\hat{\mu}(\xi)|^2 |\xi|^{-(n-\alpha)}$. If $\int |\hat{\mu}(\xi)|^2 |\xi|^{-(n-\alpha)}d\xi < \infty$ then the convergence is dominated since the factors $\hat{\phi}(\epsilon\xi)$ are bounded by 1, so the integrals converge to $\int |\hat{\mu}(\xi)|^2 |\xi|^{-(n-\alpha)}d\xi$. If $\int |\hat{\mu}(\xi)|^2 |\xi|^{-(n-\alpha)}d\xi = \infty$ then this remains true by Fatou's lemma. Accordingly, we can pass to the limit from (102) to obtain the proposition. $\square$

COROLLARY 8.7. *Suppose μ is a compactly supported probability measure on $\mathbb{R}^n$ with*

$$
\text{(103)} \qquad |\hat{\mu}(\xi)| \leq C|\xi|^{-\beta}
$$

for some $0 < \beta < n/2$, or more generally that (103) is true in the sense of L^2 means:

$$
\text{(104)} \qquad \int_{D(0,N)} |\hat{\mu}(\xi)|^2 d\xi \leq CN^{n-2\beta}.
$$

Then the dimension of the support of μ is at least 2β.

PROOF. It suffices by Proposition 8.4 to show that if (104) holds then $I_\alpha(\mu) < \infty$ for all $\alpha < 2\beta$. However,

$$\int_{|\xi| \geq 1} |\hat{\mu}(\xi)|^2 |\xi|^{-(n-\alpha)} d\xi \lesssim \sum_{j=0}^{\infty} 2^{-j(n-\alpha)} \int_{2^j \leq |\xi| \leq 2^{j+1}} |\hat{\mu}(\xi)|^2 d\xi$$

$$\lesssim \sum_{j=0}^{\infty} 2^{-j(n-\alpha)} 2^{j(n-2\beta)}$$

$$< \infty$$

if $\alpha < 2\beta$ and (104) holds. Observe also that the integral over $|\xi| \leq 1$ is finite since $|\hat{\mu}(\xi)| \leq \|\mu\| = 1$. This completes the proof in view of Proposition 8.5. $\square$

One can ask the converse question, whether a compact set with dimension α must support a measure μ with

$$(105) \qquad\qquad |\hat{\mu}(\xi)| \leq C_\epsilon (1 + |\xi|)^{-\frac{\alpha}{2}+\epsilon}$$

for all $\epsilon > 0$. The answer is (emphatically) no[3]. Indeed, there are many sets with positive dimension which do not support any measure whose Fourier transform goes to zero as $|\xi| \to \infty$. The easiest way to see this is to consider the line segment $E = [0,1] \times \{0\} \subset \mathbb{R}^2$. E has dimension 1, but if μ is a measure supported on E then $\hat{\mu}(\xi)$ depends on ξ_1 only, so it cannot go to zero at ∞. If one considers only the case $n = 1$, this question is related to the classical question of "sets of uniqueness". See e.g. [**28**], [**40**]. One can show for example that the standard $\frac{1}{3}$ Cantor set does not support any measure such that $\hat{\mu}$ vanishes at ∞.

Indeed, it is nontrivial to show that a "noncounterexample" exists, i.e. a set E with given dimension α which supports a measure satisfying (105). We describe a construction of such a set due to R. Kaufman in the next section.

As a typical application (which is also important in its own right) we now discuss a special case of Marstrand's projection theorem. Let e be a unit vector in $\mathbb{R}^n$ and $E \subset \mathbb{R}^n$ a compact set. The projection $P_e(E)$ is the set $\{x \cdot e : x \in E\}$. We want to relate the dimensions of E and of its projections. Notice first of all that $\dim P_e E \leq \dim E$; this follows from the definition of dimension and the fact that the projection P_e is a Lipschitz function.

A reasonable example, although not very typical, is a smooth curve in $\mathbb{R}^2$. This is one-dimensional, and most of its projection will be also one-dimensional. However, if the curve is a line, then one of its projections will be just a point.

THEOREM 8.8. (Marstrand's projection theorem for 1-dimensional projections) *Assume that $E \subset \mathbb{R}^n$ is compact and $\dim E = \alpha$. Then*

(i) If $\alpha \leq 1$ then for a.e. $e \in S^{n-1}$ we have $dim P_e E = \alpha$.

---

[3]On the other hand, if one interprets decay in an L^2 averaged sense the answer becomes yes, because the calculation in the proof of the above corollary is reversible.

(ii) If $\alpha > 1$ then for a.e. $e \in S^{n-1}$ the projection $P_e E$ has positive one-dimensional Lebesgue measure.

PROOF. If μ is a measure supported on E, $e \in S^{n-1}$, then the projected measure μ_e is the measure on $\mathbb{R}$ defined by

$$\int f d\mu_e = \int f(x \cdot e) d\mu(x)$$

for continuous f. Notice that $\widehat{\mu_e}$ may readily be calculated from this definition:

$$\widehat{\mu_e}(k) = \int e^{-2\pi i k x \cdot e} d\mu(x)$$
$$= \hat{\mu}(ke).$$

Let $\alpha < \dim E$, and let μ be a measure supported on E with $I_\alpha(\mu) < \infty$. We have then

(106) $$\int |\hat{\mu}(ke)|^2 |k|^{-1+\alpha} dk d\sigma(e) < \infty$$

by Proposition 8.5 and polar coordinates. Thus, for a.e. e we have

$$\int |\hat{\mu}(ke)|^2 |k|^{-1+\alpha} dk < \infty.$$

It follows by Proposition 8.5 with $n = 1$ that for a.e. e the projected measure μ_e has finite α-dimensional energy. This and Proposition 8.4 give part (i), since μ_e is supported on the projected set $P_e E$. For part (ii), we note that if $\dim E > 1$ we can take $\alpha = 1$ in (106). Thus $\widehat{\mu_e}$ is in L^2 for almost all e. By Theorem 3.13, this condition implies that μ_e has an L^2 density, and in particular is absolutely continuous with respect to Lebesgue measure. Accordingly $P_e E$ must have positive Lebesgue measure. $\qquad\square$

REMARK Theorem 8.8 has a natural generalization to k-dimensional instead of 1-dimensional projections, which is proved in the same way. See [**10**].

CHAPTER 9

Sets with Maximal Fourier Dimension and Distance Sets

A. Sets with maximal Fourier dimension

Jarnik's theorem is the following Proposition 9.A.1. Fix a number $\alpha > 0$, and let

$$E_\alpha = \{x \in \mathbb{R} : \exists \text{ infinitely many rationals } \frac{a}{q} \text{ such that } |x - \frac{a}{q}| \leq q^{-(2+\alpha)}\}.$$

PROPOSITION 9.A.1. *The Hausdorff dimension of E_α is equal to $\frac{2}{2+\alpha}$.*

PROOF. We show only that $\dim E_\alpha \leq \frac{2}{2+\alpha}$. The converse inequality is not much harder (see [10]) but we have no need to give a proof of it since it follows from Theorem 9.A.2 below using Corollary 8.7.

It suffices to prove the upper bound for $E_\alpha \cap [-N, N]$. Consider the set of intervals $I_{aq} = (\frac{a}{q} - q^{-(2+\alpha)}, \frac{a}{q} + q^{-(2+\alpha)})$, where $0 \leq a \leq Nq$ are integers. Then

$$\sum_{q > q_0} \sum_a |I_{aq}|^\beta \approx \sum_{q > q_0} q \cdot q^{-\beta(2+\alpha)},$$

which is finite and goes to 0 as $q_0 \to \infty$ if $\beta > \frac{2}{2+\alpha}$. For any given q_0 the set $\{I_{aq} : q > q_0\}$ covers $E_\alpha \cap [-N, N]$, which therefore has $H_\beta(E_\alpha \cap [-N, N]) = 0$ when $\beta > \frac{2}{2+\alpha}$, as claimed. $\qquad\square$

THEOREM 9.A.2 (Kaufman [21]). *For any $\alpha > 0$ there is a positive measure μ supported on a subset of E_α such that*

$$(107) \qquad |\hat{\mu}(\xi)| \leq C_\epsilon |\xi|^{-\frac{1}{2+\alpha} + \epsilon}$$

for all $\epsilon > 0$.

This shows then that Corollary 8.7 is best possible of its type.

The proof is most naturally done using periodic functions, so we start with the following general remarks concerning "periodization" and "deperiodization". Let $\mathbb{T}^n$ be the n-torus which we regard as $[0,1] \times \ldots \times [0,1]$ with edges identified; thus a function on $\mathbb{T}^n$ is the same as a function on $\mathbb{R}^n$ periodic for the lattice $\mathbb{Z}^n$.

If $f \in L^1(\mathbb{T}^n)$ then one defines its Fourier coefficients by

$$\hat{f}(k) = \int_{\mathbb{T}^n} f(x) e^{-2\pi i k \cdot x} dx, \quad k \in \mathbb{Z}^n$$

67

and one also makes the analogous definition for measures. If f is smooth then one has

$$\tag{108} |\hat{f}(k)| \leq C_N |k|^{-N} \text{ for all } N$$

and $\sum_{k \in \mathbb{Z}^n} \hat{f}(k) e^{2\pi i k \cdot x} = f(x)$.

Also, if $f \in L^1(\mathbb{R}^n)$ one defines its periodization by

$$f_{per}(x) = \sum_{\nu \in \mathbb{Z}^n} f(x - \nu).$$

Then $f_{per} \in L^1(\mathbb{T}^n)$, and we have

LEMMA 9.A.3 *If $k \in \mathbb{Z}^n$ then $\widehat{f_{per}}(k) = \hat{f}(k)$.*

PROOF.

$$\begin{aligned}
\hat{f}(k) &= \int_{\mathbb{R}^n} f(x) e^{-2\pi i k \cdot x} dx \\
&= \sum_{\nu \in \mathbb{Z}^n} \int_{[0,1] \times \ldots \times [0,1] + \nu} f(x) e^{-2\pi i k \cdot x} dx \\
&= \sum_{\nu \in \mathbb{Z}^n} \int_{[0,1] \times \ldots \times [0,1]} f(x - \nu) e^{-2\pi i k \cdot (x - \nu)} dx \\
&= \int_{[0,1] \times \ldots \times [0,1]} \sum_{\nu \in \mathbb{Z}^n} f(x - \nu) e^{-2\pi i k \cdot (x - \nu)} dx \\
&= \int_{[0,1] \times \ldots \times [0,1]} f_{per}(x) e^{-2\pi i k \cdot x} dx.
\end{aligned}$$

At the last line we used that $e^{-2\pi i k \cdot \nu} = 1$. $\qquad\square$

Suppose now that f is a smooth function on $\mathbb{T}^n$; regard it as a periodic function on $\mathbb{R}^n$. Let $\phi \in \mathcal{S}$ and consider the function $F(x) = \phi(x) f(x)$. We have then

$$\begin{aligned}
\hat{F}(\xi) &= \sum_{\nu \in \mathbb{Z}^n} \hat{f}(\nu) \int e^{-2\pi i (\xi - \nu) \cdot x} \phi(x) dx \\
&= \sum_{\nu \in \mathbb{Z}^n} \hat{f}(\nu) \hat{\phi}(\xi - \nu).
\end{aligned}$$

This formula extends by a limiting argument to the case where the smooth function f is replaced by a measure; we omit this argument[1]. Thus we have the following: let μ be a measure on $\mathbb{T}^n$, let $\phi \in \mathcal{S}$, and define a measure ν on $\mathbb{R}^n$ by

$$\tag{109} d\nu(x) = \phi(x) d\mu(\{x\}),$$

---

[1]It is based on the fact that every measure on $\mathbb{T}^n$ is the weak∗ limit of a sequence of absolutely continuous measures with smooth densities, which is a corollary e.g. of the Stone-Weierstrass theorem.

where $\{x\}$ is the fractional part of x. Then for $\xi \in \mathbb{R}^n$

$$(110) \qquad \hat{\nu}(\xi) = \sum_{k \in \mathbb{Z}^n} \hat{\mu}(k)\hat{\phi}(\xi - k).$$

A corollary of this formula by simple estimates with absolutely convergent sums, using the Schwartz decay of $\hat{\phi}$, is the following.

LEMMA 9.A.4 *If μ is a measure on $\mathbb{T}^n$, satisfying*

$$|\hat{\mu}(k)| \le C(1 + |k|)^{-\alpha}$$

for a certain $\alpha > 0$, and if $\nu \in M(\mathbb{R}^n)$ is defined by (109), then

$$|\hat{\nu}(\xi)| \le C'(1 + |\xi|)^{-\alpha}.$$

This can be proved by using (110) and considering the range $|\xi - k| \le |\xi|/2$ and its complement separately. The details are left to the reader.

We now start to construct a measure on the 1-torus $\mathbb{T}$, which will be used to prove Theorem 9.A.2.

Let ϕ be a nonnegative C_0^∞ function on $\mathbb{R}$ supported in $[-1, 1]$ and with $\int \phi = 1$. Define $\phi^\epsilon(x) = \epsilon^{-1}\phi(\epsilon^{-1}x)$ and let Φ^ϵ be the periodization of ϕ^ϵ.

Let $\mathcal{P}(M)$ be the set of prime numbers which lie in the interval $(\frac{M}{2}, M]$. By the prime number theorem, $|\mathcal{P}(M)| \approx \frac{M}{\log M}$ for large M. If $p \in \mathcal{P}(M)$ then the function $\Phi_p^\epsilon(x) \overset{def}{=} \Phi^\epsilon(px)$ is again 1-periodic[2] and we have

$$(111) \qquad \widehat{\Phi_p^\epsilon}(k) = \begin{cases} \hat{\phi}(\epsilon\frac{k}{p}) & \text{if } p \mid k, \\ 0 & \text{otherwise.} \end{cases}$$

To see this, start from the formula

$$\widehat{\Phi^\epsilon}(k) = \widehat{\phi^\epsilon}(k) = \hat{\phi}(\epsilon k)$$

which follows from Lemma 9.A.3. Thus

$$\Phi_\epsilon(x) = \sum_k \hat{\phi}(\epsilon k)e^{2\pi i k \cdot x},$$

$$\Phi_p^\epsilon(x) = \sum_k \hat{\phi}(\epsilon k)e^{2\pi i k p \cdot x},$$

which is equivalent to (111).

Now define

$$F = \frac{1}{|\mathcal{P}(M)|} \sum_{p \in \mathcal{P}(M)} \Phi_p^\epsilon.$$

Then F is smooth, 1-periodic, and $\int_0^1 F = 1$ (cf. (112)). Of course F depends on ϵ and M but we suppress this dependence.

LEMMA The Fourier coefficients of F behave as follows:

$$(112) \qquad\qquad\qquad \hat{F}(0) = 1,$$

---

[2] Of course for fixed p it is p^{-1}-periodic

$$(113) \qquad \hat{F}(k) = 0 \text{ if } 0 < |k| \le \tfrac{M}{2},$$

for any N there is C_N such that

$$(114) \qquad |\hat{F}(k)| \le C_N \frac{\log |k|}{M}\left(1 + \frac{\epsilon |k|}{M}\right)^{-N} \text{ for all } k \ne 0.$$

PROOF. Both (112) and (113) are selfevident from (111). For (114) we use that a given integer $m > 0$ has at most $C\frac{\log m}{\log M}$ different prime divisors in the interval $(M/2, M]$. Hence, by (111) and the Schwartz decay of $\hat{\phi}$,

$$|\hat{F}(k)| \le \frac{\log M}{M} \cdot \frac{\log |k|}{\log M} \cdot C_N\left(1 + \frac{\epsilon |k|}{M}\right)^{-N}$$

as claimed. $\qquad\square$

We now make up our mind to choose $\epsilon = M^{-(1+\alpha)}$, and denote the function F by F_M. Thus we have the following

$$(115) \qquad \mathrm{supp} F_M \subset \{x : \; |x - \tfrac{a}{p}| \le p^{-(2+\alpha)} \text{ for some } p \in \mathcal{P}_M, a \in [0, p]\},$$

$$(116) \qquad |\widehat{F_M}(k)| \le C_N \frac{\log |k|}{M}\left(1 + \frac{|k|}{M^{2+\alpha}}\right)^{-N},$$

and furthermore F_M is nonnegative and satisfies (112) and (113).

It is easy to deduce from (116) with $N = 1$ that

$$(117) \qquad |\widehat{F_M}(k)| \lesssim |k|^{-\frac{1}{2+|\alpha|}} \log |k|$$

uniformly in M. In view of (115) we could now try to prove the theorem by taking a weak limit of the measures $F_M dx$ and then using Lemma 9.A.4. However, this would not be correct, since the set E_α is not closed and ((115) notwithstanding) there is no reason why the weak limit should be supported on E_α. Indeed, (113) and (112) imply easily that the weak limit is the Lebesgue measure. The following is the standard way of getting around this kind of problem. It has something in common with the classical "Riesz product" construction; see [25]. Essentially, to get the right support properties we could take the weak limit of the measures $F_{M_1} \ldots F_{M_N}$ for some sequence $M_N \to \infty$, but then we would lose the estimate (117). This can be overcome by observing that (117) is sharp only for a relatively small range of k, different for each M, so that we can get better estimates (and preserve (117) in the end) if we replace the functions F_M by averages of many such functions and then take the limit of the products.

In what follows, the constants C, C_1, etc. may change from line to line but will always be independent of k, l, M.

I. First consider a fixed smooth function ψ on $\mathbb{T}$. We claim that if M has been chosen large enough then

$$(118) \qquad |\widehat{\psi F_M}(k) - \hat{\psi}(k)| \le \begin{cases} C\frac{\log |k|}{M}\left(1 + \frac{|k|}{M^{2+\alpha}}\right)^{-100} & \text{when } |k| \ge \frac{M}{4} \\[2mm] CM^{-100} & \text{when } |k| \le \frac{M}{4}. \end{cases}$$

Indeed, suppose first that $|k| \leq \frac{M}{4}$. We write using (110) and (113), (112)

$$
(119)
$$
$$
\begin{aligned}
|\widehat{\psi F_M}(k) - \widehat{\psi}(k)| &= |\textstyle\sum_{l \in \mathbb{Z}} \widehat{\psi}(l)\widehat{F_M}(k-l) - \widehat{\psi}(k)| \\
&= |\textstyle\sum_{l:|k-l| \geq M/2} \widehat{\psi}(l)\widehat{F_M}(k-l)| \\
&\leq C_1 M^{-100} \max_{l:|k-l| \geq M/2} |\widehat{F_M}(k-l)| \\
&\leq C_2 M^{-100} \max_{l:|k-l| \geq M/2} \frac{\log|k-l|}{M}\left(1 + \frac{|k-l|}{M^{2+\alpha}}\right)^{-200}.
\end{aligned}
$$

On the next to last line we used that $|l| \geq M/4$ when $|k| \leq \frac{M}{4}$ and $|k-l| \geq M/2$, so that by (108) $\sum_{l:|k-l| \geq M/2} |\widehat{\psi}(l)| \lesssim M^{-N}$ for any N. The last line followed from (116). Note that if M has been chosen large enough, the function

$$
f(t) = \frac{\log t}{M}\left(1 + \frac{t}{M^{2+\alpha}}\right)^{-N}
$$

is decreasing for $t \geq M/10$. It follows that

$$
\begin{aligned}
|\widehat{\psi F_M}(k) - \widehat{\psi}(k)| &\leq C_3 M^{-100}\frac{\log(M/2)}{M}\left(1 + \frac{M/2}{M^{2+\alpha}}\right)^{-200} \\
&\leq CM^{-100},
\end{aligned}
$$

which proves the second part of (118). To prove the first part, we start as in (119) and consider separately the range $|k-l| \geq |k|/2$ and its complement. For $|k-l| \geq |k|/2$ we use the same argument as above, with the lower bound $M/2$ on $|k-l|$ replaced by $|k|/2$. For $|k-l| \leq |k|/2$ we have $|l| \geq |k|/2$, hence the estimate follows easily from (108) and the fact that $|\widehat{F_M}(k)| \leq 1$. The details are left to the reader.

II. Let

$$
g(r) = \begin{cases}
r^{-\frac{1}{2+\alpha}} \log r & \text{when } r \geq r_0, \\[2ex]
r_0^{-\frac{1}{2+\alpha}} \log r_0 & \text{when } r \leq r_0,
\end{cases}
$$

where $r_0 > 1$ is chosen so that $g(r) \leq 1$ and $g(r)$ is nonincreasing for all r. Then for any $\psi \in C^\infty(\mathbb{T})$, $\epsilon > 0$, and $M_0 > 10r_0$ we can choose N large enough and a rapidly increasing sequence $M_0 < M_1 < M_2 < \cdots < M_N$ so that

$$
(120) \qquad\qquad |\widehat{\psi G}(k) - \widehat{\psi}(k)| \leq \epsilon g(|k|),
$$

where $G = N^{-1}(F_{M_1} + \cdots + F_{M_N})$.

This can be done as follows. Fix N and $\tilde{M}$ sufficiently large so that $\frac{C}{N} < \frac{\epsilon}{100}$, $\tilde{M} \geq M_0$, and

$$
(121) \qquad\qquad CM^{-100} < \frac{\epsilon}{100}g(M) \text{ if } M \geq \tilde{M}.
$$

Here C is a large constant which we will later fix to be equal to that in (125). We now choose $M_1, M_2, \ldots, M_N$ inductively so that (121) holds,

$M_{j+1} > 4M_j$ and

$$(122) \qquad \frac{1}{N} \sum_{i=1}^{j} E_i(k) \leq \frac{\epsilon}{100} g(|k|) \text{ if } |k| > M_{j+1},$$

where $E_i(k) = |\widehat{\psi F_{M_i}}(k) - \widehat{\psi}(k)|$. This is possible since by (118) $|E_i(k)|$ vanishes at infinity much faster than $g(|k|)$ for each fixed i.

We claim that (120) holds for this choice of M_j. To show this, we start with

$$(123) \qquad |\widehat{\psi G}(k) - \widehat{\psi}(k)| \leq \frac{1}{N} \sum_{i=1}^{N} E_i(k).$$

Assume that $M_j \leq |k| \leq M_{j+1}$ (the cases $|k| \leq M_1$ and $|k| \geq M_N$ are similar and are left to the reader). By (122) we have

$$(124) \qquad \frac{1}{N} \sum_{i=1}^{j-1} E_i(k) \leq \frac{\epsilon}{100} g(|k|).$$

To estimate the remaining terms we use the following easy consequence of (118):

$$(125) \qquad E_j(k) \leq \begin{cases} Cg(|k|) & \text{if } |k| \geq M_j/4, \\ CM_j^{-100} & \text{if } |k| \leq M_j/4, \end{cases}$$

with the constant C independent of k, j. (The second part is immediate from (118), and the first part is a simple exercise in calculus.) Thus

$$(126) \qquad \begin{aligned} \sum_{i=j}^{N} \frac{1}{N} E_i(k) &\leq 2\frac{C}{N} g(|k|) + \sum_{i=j+1}^{N} \frac{C}{N} M_i^{-100} \\ &\leq \frac{\epsilon}{100} g(|k|) + \sum_{i=j+1}^{N} \frac{C}{N} M_i^{-100}. \end{aligned}$$

We used the first part of (125) to estimate the term with $i = j$, the second part to estimate the terms with $i \geq j + 2$, and estimated the term with $i = j + 1$ by the sum of the two bounds.

Combining (124) and (126) we obtain that

$$\begin{aligned} |\widehat{\psi G}(k) - \widehat{\psi}(k)| &\leq \frac{3\epsilon}{100} g(|k|) + \frac{C}{N} \sum_{i=j+1}^{N} M_i^{-100} \\ &\leq \frac{\epsilon}{100N} \sum_{i=j+1}^{N} g(M_i) \\ &\leq \frac{\epsilon}{100} g(|k|), \end{aligned}$$

which proves (120). For the last two inequalities we used (121) and that $g(r)$ is nonincreasing.

We note that the support properties of G are similar to those of the F's. Namely, it follows from (115) that

$$(127) \quad \operatorname{supp} G \subset \{x : |x - \frac{a}{p}| \leq p^{-(2+\alpha)} \text{ for some } p \in (\frac{M_1}{2}, M_N), \ a \in [0, p]\}.$$

III. We now construct inductively the functions G_m and H_m, $m = 1, 2, \ldots$, as follows. Let $G_0 \equiv 1$. If G_m has been constructed, we let G_{m+1} be as in step II with $\psi = G_0 G_1 \ldots G_m$, $M_0 \geq 10 r_0 + m$, and $\epsilon = 2^{-m-2}$. Then the functions $H_m = G_1 \ldots G_m$ satisfy

$$\frac{1}{2} \leq \widehat{H_m}(0) \leq \frac{3}{2}$$

for each m and moreover the estimate (117) holds also for the H's, i.e.

$$|\widehat{H_m}(k)| \lesssim |k|^{-\frac{1}{2+\alpha}} \log |k|.$$

IV. Now let μ be a weak-$*$ limit point of the sequence $\{H_m dx\}$. The support of μ is contained in the intersection of the supports of the H_m's, hence by (127) it is a compact subset of E_α. From step III we have $|\hat{\mu}(k)| \lesssim |k|^{-\frac{1}{2+\alpha}} \log |k|$. The theorem now follows by Lemma 9.A.4. $\qquad \square$

B. Distance sets

If E is a compact set in $\mathbb{R}^2$ (or more generally in $\mathbb{R}^n$), the distance set $\Delta(E)$ is defined as

$$\Delta(E) = \{|x - y| : x, y \in E\}.$$

One version of *Falconer's distance set problem* is the conjecture that

$$E \subset \mathbb{R}^2, \ \dim E > 1 \Rightarrow |\Delta(E)| > 0.$$

One can think of this as a version of the Marstrand projection theorem where the nonlinear projection $(x, y) \to |x - y|$ replaces the linear ones. In fact, it is also possible to make the stronger conjecture that the "pinned" distance sets

$$\{|x - y| : y \in E\}$$

should have positive measure for some $x \in E$, or for a set of $x \in E$ with large Hausdorff dimension. This would be analogous to Theorem 8.8 with the nonlinear maps $y \to |x - y|$ replacing the projections P_e.

Alternately, one can consider this problem as a continuous analogue of a well known open problem in discrete geometry (Erdős' distance set problem): prove that for finite sets $F \subset \mathbb{R}^2$ there is a bound $|\Delta(F)| \geq C_\epsilon^{-1} |F|^{1-\epsilon}$, $\epsilon > 0$. The example $F = \mathbb{Z}^2 \cap D(0, N)$, $N \to \infty$ can be used to show that in Erdős' problem one cannot take $\epsilon = 0$, and a related example [11] shows that in Falconer's problem it does not suffice to assume that $H_1(E) > 0$. The current best result on Erdős' problem is $\epsilon = \frac{1}{7}$ due to Solymosi and Tóth [31] (there were many previous contributions), and on Falconer's problem the current best result is $\dim E > \frac{4}{3}$ due to myself [37] using previous work of Mattila [24] and Bourgain [4].

The strongest results on Falconer's problem have been proved using Fourier transforms in a manner analogous to the proof of Theorem 8.8. We describe the basic strategy, which is due to Mattila [24]. Given a measure μ on E, there is a natural way to put a measure on $\Delta(E)$, namely push forward the measure $\mu \times \mu$ by the map $\Delta : (x, y) \to |x - y|$. If one can show that the pushforward measure has an L^2 Fourier transform, then $\Delta(E)$ must have positive measure by Theorem 3.13.

In fact one proceeds slightly differently for technical reasons. Let μ be a measure in $\mathbb{R}^2$, then [**24**] one associates to it the measure ν defined as follows. Let $\nu_0 = \Delta(\mu \times \mu)$, i.e.

$$\int f d\nu_0 = \int f(|x - y|) d\mu(x) d\mu(y).$$

Observe that

$$\int t^{-\frac{1}{2}} d\nu_0(t) = I_{\frac{1}{2}}(\mu).$$

Thus if $I_{\frac{1}{2}}(\mu) < \infty$, as we will always assume, then the measure we now define will be in $M(\mathbb{R})$. Namely, let

$$(128) \qquad d\nu(t) = e^{i\frac{\pi}{4}} t^{-\frac{1}{2}} d\nu_0(t) + e^{-i\frac{\pi}{4}} |t|^{-\frac{1}{2}} d\nu_0(-t).$$

Since ν_0 is supported on $\Delta(E)$, ν is supported on $\Delta(E) \cup -\Delta(E)$.

PROPOSITION 9.B.1 (Mattila [**24**]) *Assume that $I_\alpha(\mu) < \infty$ for some $\alpha > 1$. Then the following are equivalent:*
(i) $\hat{\nu} \in L^2(\mathbb{R})$,
(ii) the estimate

$$(129) \qquad \int_{R=1}^{\infty} \left(\int |\hat{\mu}(Re^{i\theta})|^2 d\theta \right)^2 R \, dR < \infty.$$

COROLLARY 9.B.2 [**24**] *Suppose that $\alpha > 1$ is a number with the following property: if μ is a positive compactly supported measure with $I_\alpha(\mu) < \infty$ then*

$$(130) \qquad \int |\hat{\mu}(Re^{i\theta})|^2 d\theta \leq C_\mu R^{-(2-\alpha)}.$$

Then any compact subset of $\mathbb{R}^2$ with dimension $> \alpha$ must have a positive measure distance set.

Here and below we identify $\mathbb{R}^2$ with $\mathbb{C}$ in the obvious way.

PROOF OF THE COROLLARY. Assume $\dim E > \alpha$. Then E supports a measure with $I_\alpha(\mu) < \infty$. We have

$$\int_{R=1}^{\infty} \left(\int |\hat{\mu}(Re^{i\theta})|^2 d\theta \right)^2 R \, dR \; \lesssim \; \int_{R=1}^{\infty} \left(\int |\hat{\mu}(Re^{i\theta})|^2 d\theta \right) R^{-(2-\alpha)} R \, dR$$

$$< \; \infty.$$

On the first line we used (130) to estimate one of the two angular integrals, and the last line then follows by recognizing that the resulting expression corresponds to the Fourier representation of the energy in Proposition 8.5. By Proposition 9.B.1 $\Delta(E) \cup -\Delta(E)$ supports a measure whose Fourier transform is in L^2, which suffices by Theorem 3.13. $\square$.

At the end of the section we will prove (130) in the easy case $\alpha = \frac{3}{2}$ where it follows from the uncertainty principle; we believe this is due to P.

Sjölin. It is known [**37**] that (130) holds when $\alpha > \frac{4}{3}$, and this is essentially sharp since (130) fails when $\alpha < \frac{4}{3}$. The negative result follows from a variant on the Knapp argument (Remark 4. at the beginning of Chapter 7); this is due to [**24**], and is presented also in several other places, e.g. [**37**]. The positive result requires more sophisticated L^p type arguments.

Before proving the proposition we record a few more formulas. Let σ_R be the angular measure on the circle of radius R centered at zero; thus we are normalizing the arclength measure on this circle to have total mass 2π. Let μ be any measure with compact support. We then have

$$(131) \qquad \int |\hat{\mu}(Re^{i\theta})|^2 d\theta = \int \widehat{\sigma_R} * \mu \, d\mu.$$

This is just one more instance of the formula which first appeared in Lemma 7.2 and was used in the proof of Proposition 8.5. This version is contained in Lemma 7.2 if μ has a Schwartz space density, and a limiting argument like the one in the proof of Proposition 8.4 shows that it holds for general μ. We also record the asymptotics for $\widehat{\sigma_R}$ which of course follow from those for $\widehat{\sigma_1}$ (Corollary 6.7) using dilations. Notice that the passage from σ_1 to σ_R preserves the total mass, i.e. essentially $\sigma_R = (\sigma_1)^R$. We conclude that

$$(132) \qquad \widehat{\sigma_R}(x) = 2(R|x|)^{-\frac{1}{2}} \cos(2\pi(R|x| - \frac{1}{8})) + \mathcal{O}((R|x|)^{-\frac{3}{2}})$$

when $R|x| \geq 1$, and $|\widehat{\sigma_R}|$ is clearly also bounded independently of R.

PROOF OF THE PROPOSITION. From the definition of ν we have

$$\begin{aligned}
\hat{\nu}(k) &= e^{i\frac{\pi}{4}} \int |x-y|^{-\frac{1}{2}} e^{-2\pi i k |x-y|} d\mu(x) d\mu(y) \\
&+ e^{-i\frac{\pi}{4}} \int |x-y|^{-\frac{1}{2}} e^{2\pi i k |x-y|} d\mu(x) d\mu(y) \\
&= 2 \int |x-y|^{-\frac{1}{2}} \cos(2\pi(|k||x-y| - \frac{1}{8})) d\mu(x) d\mu(y).
\end{aligned}$$

On the other hand, by (131) and (132) we have

$$\begin{aligned}
\int |\hat{\mu}(ke^{i\theta})|^2 d\theta &= |k|^{-\frac{1}{2}} \int 2|x-y|^{-\frac{1}{2}} \cos(2\pi(|k||x-y| - \frac{1}{8})) d\mu(x) d\mu(y) \\
&+ \mathcal{O}\left(\int_{|x-y| \geq |k|^{-1}} (|k||x-y|)^{-\frac{3}{2}} d\mu(x) d\mu(y) \right) \\
&+ \mathcal{O}\left(\int_{|x-y| \leq |k|^{-1}} (|k||x-y|)^{-\frac{1}{2}} d\mu(x) d\mu(y) \right).
\end{aligned}$$

The last error term arises by comparing $\widehat{\sigma_R}$, which is bounded, to the main term on the right side of (132), which is $\mathcal{O}((R|x|)^{-\frac{1}{2}})$, in the regime $R|x| < 1$.

We may combine the two error terms to obtain

$$\int |\hat{\mu}(ke^{i\theta})|^2 d\theta = |k|^{-\frac{1}{2}} \int 2|x - y|^{-\frac{1}{2}} \cos(2\pi(|k||x - y| - \frac{1}{8}))d\mu(x)d\mu(y)$$

$$+\mathcal{O}(\int (|k||x - y|)^{-\alpha}d\mu(x)d\mu(y))$$

for any $\alpha \in [\frac{1}{2}, \frac{3}{2}]$. Therefore

$$\hat{\nu}(k) = |k|^{\frac{1}{2}} \int |\hat{\mu}(ke^{i\theta})|^2 d\theta + \mathcal{O}(|k|^{\frac{1}{2}-\alpha}I_\alpha(\mu)).$$

The error term here is evidently bounded by $|k|^{\frac{1}{2}-\alpha}I_\alpha(\mu)$ for any $\alpha \in (1, \frac{3}{2})$, and therefore belongs to $L^2(|k| \geq 1)$. We conclude then that $\hat{\nu}$ belongs to L^2 on $|k| \geq 1$ if and only if $|k|^{\frac{1}{2}} \int |\hat{\mu}(ke^{i\theta})|^2 d\theta$ does. This gives the proposition, since $\hat{\nu}$ (being the Fourier transform of a measure) clearly belongs to L^2 on $[-1, 1]$. $\qquad\square$

PROPOSITION 9.B.3 *If $\alpha \geq 1$ and if μ is a positive measure with compact support[3] then*

$$\int |\hat{\mu}(Re^{i\theta})|^2 d\theta \lesssim I_\alpha(\mu)R^{-(\alpha-1)},$$

where the implicit constant depends on a bound for the radius of a disc centered at 0 which contains the support of μ. In particular, (130) holds if $\alpha = \frac{3}{2}$.

COROLLARY 9.B.4 (originally due to Falconer [11] with a different proof) *If $\dim E > \frac{3}{2}$ then the distance set of E has positive measure.*

PROOFS. The corollary is immediate from the proposition and Corollary 9.B.2. The proof of the proposition is very similar to the proofs of Bernstein's inequality and of Theorem 7.4. We can evidently assume that R is large. Let ϕ be a radial C_0^∞ function whose Fourier transform is ≥ 1 on the support of μ. Let $d\nu(x) = (\hat{\phi}(x))^{-1}d\mu(x)$. Then it is obvious (from the definition, not the Fourier representation) that $I_\alpha(\nu) \leq I_\alpha(\mu)$. Also $\hat{\mu} = \phi * \hat{\nu}$. Accordingly

$$\int |\hat{\mu}(Re^{i\theta})|^2 d\theta = \int |\phi * \hat{\nu}(Re^{i\theta})|^2 d\theta$$

$$\lesssim \int |\phi(Re^{i\theta} - x)||\hat{\nu}(x)|^2 dx d\theta$$

$$= \int |\hat{\nu}(x)|^2 \int |\phi(x - Re^{i\theta})|d\theta dx$$

$$\lesssim R^{-1} \int_{||x|-R|\leq C} |\hat{\nu}(x)|^2 dx$$

$$\lesssim R^{-1+2-\alpha} \int |x|^{-(2-\alpha)}|\hat{\nu}(x)|^2 dx$$

$$\approx R^{1-\alpha}I_\alpha(\mu).$$

---

[3]Here, as opposed to in some previous situations, the compact support is important.

Here the second line follows by writing

$$|\phi * \hat{\nu}(Re^{i\theta})| \leq \int \sqrt{|\phi(Re^{i\theta} - x)|} \cdot \sqrt{|\phi(Re^{i\theta} - x)||\hat{\nu}(x)|} dx$$

and applying the Schwartz inequality. The fourth line follows since for fixed x the set of θ where $\phi(x - Re^{i\theta}) \neq 0$ has measure $\lesssim R^{-1}$, and is empty if $|x| - R$ is large. The proof is complete. $\qquad\square$

REMARK The exponent $\alpha - 1$ is of course far from sharp; the sharp exponent is $\frac{\alpha}{2}$ if $\alpha > 1$, $\frac{1}{2}$ if $\alpha \in [\frac{1}{2}, 1]$ and α if $\alpha < \frac{1}{2}$.

EXERCISE Prove this in the case $\alpha \leq 1$. (This is a fairly hard exercise.)

EXERCISE Carry out Mattila's construction (formula (128) and the preceding discussion) in the case where μ is a measure in $\mathbb{R}^n$ instead of $\mathbb{R}^2$, and prove analogues of Proposition 9.A.1, Corollary 9.A.2, Proposition 9.A.3. Conclude that a set in $\mathbb{R}^n$ with dimension greater than $\frac{n+1}{2}$ has a positive measure distance set. (See [24]. The dimension result is also due originally to Falconer. The conjectured sharp exponent is $\frac{n}{2}$.)

CHAPTER 10

The Kakeya Problem

A *Besicovitch set*, or a *Kakeya set*, is a compact set $E \subset \mathbb{R}^n$ which contains a unit line segment in every direction, i.e.

$$(133) \qquad \forall e \in S^{n-1} \ \exists x \in \mathbb{R}^n : \ x + te \in E \ \forall t \in [-\frac{1}{2}, \frac{1}{2}].$$

THEOREM 10.1 (Besicovitch, 1920). *If $n \geq 2$, then there are Kakeya sets in $\mathbb{R}^n$ with measure zero.*

There are many variants on Besicovitch's construction in the literature, cf. [10], Chapter 7, or [39], Chapter 1.

There is a basic open question about Besicovitch sets which can be stated vaguely as "How small can this really be?" This can be formulated more precisely in terms of fractal dimension. If one uses the Hausdorff dimension, then the main question is the following.

OPEN QUESTION: THE KAKEYA CONJECTURE. If $E \subset \mathbb{R}^n$ is a Kakeya set, does E necessarily have Hausdorff dimension n?

If $n = 2$ then the answer is yes; this was proved by Davies [8] in 1971. For general n, what is known at present is that $\dim(E) \geq \min(\frac{n+2}{2}, (2 - \sqrt{2})(n - 4) + 3)$; the first bound which is better for $n = 3$ is due to myself [38], and the second one is due to Katz and Tao [19]. Instead of the Hausdorff dimension one can use other notions of dimension, for example the Minkowski dimension defined in Chapter 8. The current best results for the upper Minkowski dimension are due to Katz, Laba and Tao [18], [22], [19].

There is also a more quantitative formulation of the problem in terms of the *Kakeya maximal functions*, which are defined as follows. For any $\delta > 0$, $e \in S^{n-1}$ and $a \in \mathbb{R}^n$, let

$$T_e^\delta(a) = \{x \in \mathbb{R}^n : |(x - a) \cdot e| \leq \frac{1}{2}, |(x - a)^\perp| \leq \delta\},$$

where $x^\perp = x - (x \cdot e)e$. Thus $T_e^\delta(a)$ is essentially the δ-neighborhood of the unit line segment in the e direction centered at a. Then the Kakeya maximal function of $f \in L_{loc}^1(\mathbb{R}^n)$ is the function $f_\delta^* : S^{n-1} \to \mathbb{R}$ defined by

$$(134) \qquad f_\delta^*(e) = \sup_{a \in \mathbb{R}^n} \frac{1}{|T_e^\delta(a)|} \int_{T_e^\delta(a)} |f|.$$

The issue is to prove a "$\delta^{-\epsilon}$" estimate for f_δ^*, i.e. an estimate of the form

$$(135) \qquad \forall \epsilon \exists C_\varepsilon : \|f_\delta^*\|_{L^p(S^{n-1})} \leq C_\varepsilon \delta^{-\varepsilon}\|f\|_p$$

for some $p < \infty$.

REMARKS 1. It is clear from the definition that

$$(136) \qquad \|f_\delta^*\|_\infty \leq \|f\|_\infty,$$

$$(137) \qquad \|f_\delta^*\|_\infty \leq \delta^{-(n-1)}\|f\|_1.$$

2. If $n \geq 2$ and $p < \infty$, there can be no bound of the form

$$(138) \qquad \|f_\delta^*\|_q \leq C\|f\|_p,$$

with C independent of δ. This can be seen as follows. Consider a zero measure Kakeya set E. Let E_δ be the δ-neighborhood of E, and let $f = \chi_{E_\delta}$. Then $f_\delta^*(e) = 1$ for all $e \in S^{n-1}$, so that $\|f_\delta^*\|_q \approx 1$. On the other hand, $\lim_{\delta \to 0} |E_\delta| = 0$, hence $\lim_{\delta \to 0}\|f_\delta\|_p = 0$ for any $p < \infty$.

3. Let $f = \chi_{D(0,\delta)}$. Then for all $e \in S^{n-1}$ the tube $T_e^\delta(0)$ contains $D(0,\delta)$, so that $f_\delta^*(e) = \frac{|D(0,\delta)|}{|T_e^\delta(0)|} \gtrsim \delta$. Hence $\|f_\delta^*\|_p \approx \delta$. However, $\|f\|_p \approx \delta^{n/p}$. This shows that (135) cannot hold for any $p < n$.

OPEN PROBLEM: THE KAKEYA MAXIMAL FUNCTION CONJECTURE. Prove that (135) holds with $p = n$, i.e.

$$(139) \qquad \forall \epsilon \; \exists C_\varepsilon : \; \|f_\delta^*\|_{L^n(S^{n-1})} \leq C_\varepsilon \delta^{-\varepsilon}\|f\|_n.$$

When $n = 2$, this was proved by Córdoba [7] in a somewhat different formulation and by Bourgain [3] as stated. These results are relatively easy; from one point of view, this is because (139) is then an L^2 estimate. In higher dimensions the problem remains open. There are partial results on (139) which can be understood as follows. Interpolating between (137), which is the best possible bound on L^1, and (139) gives a family of conjectured inequalities

$$(140) \qquad \|f_\delta^*\|_q \lesssim C_\varepsilon \delta^{-\frac{n}{p}+1-\varepsilon}\|f\|_p, \quad q = q(p).$$

Note that if (140) holds for some $p_0 > 1$, it also holds for all $1 \leq p \leq p_0$ (again by interpolating with (133)). The current best results in this direction are that (140) holds with $p = \min((n + 2)/2, (4n + 3)/7)$ and a suitable q [38], [19].

PROPOSITION 10.2. *If (135) holds for some $p < \infty$, then Besicovitch sets in $\mathbb{R}^n$ have Hausdorff dimension n.*

REMARK The inequality

$$(141) \qquad |E_\delta| \geq C_\varepsilon^{-1}\delta^\varepsilon$$

for any Kakeya set E follows immediately from (135) by the same argument that was used in Remark 2 above. (141) says that Besicovitch sets in $\mathbb{R}^n$ have lower Minkowski dimension n.

PROOF OF THE PROPOSITION. Let E be a Besicovitch set. Fix a covering of E by discs $D_j = D(x_j, r_j)$; we can assume that all r_j's are $\leq 1/100$. Let

$$J_k = \{j : \ 2^{-k} \leq r_j \leq 2^{-(k-1)}\}.$$

For every $e \in S^{n-1}$, E contains a unit line segment I_e parallel to e. Let

$$S_k = \{e \in S^{n-1} : \ |I_e \cap \bigcup_{j \in J_k} D_j| \geq \frac{1}{100k^2}\}.$$

Since $\sum_k \frac{1}{100k^2} < 1$ and $\sum_k |I_e \cap \bigcup_{j \in J_k} D_j| \geq |I_e| = 1$, it follows that $\bigcup_{k=1}^{\infty} S_k = S^{n-1}$.

Let

$$f = \chi_{F_k}, \ F_k = \bigcup_{j \in J_k} D(x_j, 10r_j).$$

Then for $e \in S_k$ we have

$$|T_e^{2^{-k}}(a_e) \cap F_k| \gtrsim \frac{1}{100k^2} |T_e^{2^{-k}}(a_e)|,$$

where a_e is the midpoint of I_e. Hence

$$(142) \qquad \|f_{2^{-k}}^*\|_p \gtrsim k^{-2}\sigma(S_k)^{1/p}.$$

On the other hand, (135) implies that

$$(143) \qquad \|f_{2^{-k}}^*\|_p \leq C_\varepsilon 2^{k\varepsilon}\|f\|_p \leq C_\varepsilon 2^{k\varepsilon}(|J_k| \cdot 2^{-(k-1)np})^{1/p}.$$

Comparing (142) and (143), we see that

$$\sigma(S_k) \lesssim 2^{kp\varepsilon - kn} k^{2p}|J_k| \lesssim 2^{-k(n-2p\varepsilon)}|J_k|.$$

Therefore

$$\sum_j r_j^{n-2p\varepsilon} \geq \sum_k 2^{-k(n-2p\varepsilon)}|J_k| \gtrsim \sum_k \sigma(S_k) \gtrsim 1.$$

We have shown that $\sum r_j^\alpha \gtrsim 1$ for any $\alpha < n$, which implies the claimed Hausdorff dimension bound. $\qquad \Box$

REMARK By the same argument as in the proof of Proposition 10.2, (140) implies that the dimension of a Kakeya set in $\mathbb{R}^n$ is at least p.

A. The n = 2 case

THEOREM 10.3. *If $n = 2$, then there is a bound*

$$\|f_\delta^*\|_2 \leq C(\log \frac{1}{\delta})^{1/2}\|f\|_2.$$

We give two different proofs of the theorem. The first one is due to Bourgain [3] and uses Fourier analysis. The second one is due to Córdoba [7] and is based on geometric arguments.

PROOF 1. (Bourgain) We can assume that f is nonnegative. Let $\rho_\delta^e(x) = (2\delta)^{-1}\chi_{T_e^\delta(0)}$, then

$$f_\delta^*(e) = \sup_{a \in \mathbb{R}^2} (\rho_\delta^e * f)(a).$$

Let ψ be a nonnegative Schwartz function on $\mathbb{R}$ such that $\hat{\phi}$ has compact support and $\phi(x) \geq 1$ when $|x| \leq 1$. Define $\psi : \mathbb{R}^2 \to \mathbb{R}$ by

$$\psi(x) = \phi(x_1)\delta^{-1}\phi(\delta^{-1}x_2).$$

Note that $\psi \geq \rho_\delta^e$ when $e = e_1$, so that $f_\delta^*(e_1) \leq \sup_a(\psi * f)(a)$. Similarly

$$f_\delta^*(e) \leq \sup_a(\psi_e * f)(a),$$

where $\psi_e = \psi \circ p_e$ for an appropriate rotation p_e. Hence

$$(144) \qquad f_\delta^*(e) \leq \|\psi_e * f\|_\infty \leq \|\widehat{\psi_e}\hat{f}\|_1 = \int |\widehat{\psi_e}(\xi)| \cdot |\hat{f}(\xi)|d\xi.$$

By Hölder's inequality,

$$(145) \qquad \begin{aligned} &\int |\widehat{\psi_e}(\xi))| \, |\hat{f}(\xi)|d\xi \\ &\leq \left(\int |\widehat{\psi_e}(\xi)| \, |\hat{f}(\xi)|^2(1+|\xi|)d\xi\right)^{1/2}\left(\int \frac{|\widehat{\psi_e}(\xi)|}{1+|\xi|}d\xi\right)^{1/2}. \end{aligned}$$

Note that $\widehat{\psi_e} = \hat{\psi} \circ p_e$ and $\hat{\psi} = \hat{\phi}(x_1)\hat{\phi}(\delta x_2)$, so that $|\widehat{\psi_e}| \lesssim 1$ and $\hat{\psi}$ is supported on a rectangle R_e of size about $1 \times 1/\delta$. Accordingly,

$$(146) \qquad \int \frac{|\widehat{\psi_e}(\xi)|}{1+|\xi|}d\xi \lesssim \int_{R_e} \frac{d\xi}{1+|\xi|} \approx \int_1^{1/\delta} \frac{ds}{s} = \log(\frac{1}{\delta}).$$

Using (144), (145) and (146) we obtain

$$\begin{aligned} \|f_\delta^*\|_2^2 \; &\lesssim \log(\tfrac{1}{\delta})\left(\int |\widehat{\psi_e}(\xi)| \, |\hat{f}(\xi)|^2(1+|\xi|)d\xi\right)^{1/2} \\ &\lesssim \log(\tfrac{1}{\delta}) \int_{\mathbb{R}^2} |\hat{f}(\xi)|^2(1+|\xi|)\left(\int_{S^1} |\widehat{\psi_e}(\xi)|de\right)d\xi \\ &\lesssim \log\tfrac{1}{\delta} \int_{\mathbb{R}^2} |\hat{f}(\xi)|^2 d\xi \\ &= \log\tfrac{1}{\delta}\|f\|_2^2. \end{aligned}$$

Here the third line follows since for fixed ξ the set of $e \in S^{n-1}$ where $\widehat{\psi_e}(\xi) \neq 0$ has measure $\lesssim 1/(1+|\xi|)$. The proof is complete. $\qquad\square$

REMARK For $n \geq 3$, the same argument shows that

$$(147) \qquad \|f_\delta^*\|_2 \lesssim \delta^{-(n-2)/2}\|f\|_2,$$

which is the best possible L^2 bound.

PROOF 2. (Córdoba) The proof uses the following duality argument.

LEMMA 10.4. *Let $1 < p < \infty$, and let p' be the dual exponent of p: $\frac{1}{p} + \frac{1}{p'} = 1$. Suppose that p has the following property: if $\{e_k\} \subset S^{n-1}$ is a maximal δ-separated set, and if $\delta^{n-1}\sum_k y_k^{p'} \leq 1$, then for any choice of points $a_k \in \mathbb{R}^n$ we have*

$$\left\|\sum_k y_k \chi_{T_{e_k}^\delta(a_k)}\right\|_{p'} \leq A.$$

Then there is a bound

$$\|f_\delta^*\|_{L^p(S^{n-1})} \lesssim A\|f\|_p.$$

PROOF. Let $\{e_k\}$ be a maximal δ-separated subset of S^{n-1}. Observe that if $|e - e'| < \delta$ then $f_\delta^*(e) \le C f_\delta^*(e')$; this is because any $T_e^\delta(a)$ can be covered by a bounded number of tubes $T_{e'}^\delta(a')$. Therefore

$$
\begin{aligned}
\|f_\delta^*\|_p^p &\le \sum_k \int_{D(e_k,\delta)} |f_\delta^*(e)|^p de \\
&\lesssim \left(\delta^{n-1} \sum_k |f_\delta^*(e_k)|^p\right)^{1/p} \\
&= \delta^{n-1} \sum_k y_k |f_\delta^*(e_k)|
\end{aligned}
$$

for some sequence y_k with $\sum_k y_k^{p'} \delta^{n-1} = 1$. On the last line we used the duality between l_p and $l_{p'}$. Hence

$$\|f_\delta^*\|_p^p \lesssim \delta^{n-1} \sum_k y_k \frac{1}{|T_{e_k}^\delta(a_k)|} \int_{T_{e_k}^\delta(a_k)} |f|$$

for some choice of $\{a_k\}$. Since $|T_{e_k}^\delta(a_k)| \approx \delta^{n-1}$, it follows that

$$
\begin{aligned}
\|f_\delta^*\|_p^p &\lesssim \int \left(\sum_k y_k \chi_{T_{e_k}^\delta(a_k)}\right) |f| \\
&\le \left\|\sum_k y_k \chi_{T_{e_k}^\delta(a_k)}\right\|_{p'} \cdot \|f\|_p \quad \text{(Hölder's inequality)} \\
&\le A\|f\|_p
\end{aligned}
$$

as claimed. $\square$

We continue with Córdoba's proof. In view of Lemma 10.4, it suffices to prove that for any sequence $\{y_k\}$ with $\delta \sum y_k^2 = 1$ and any maximal δ-separated subset $\{e_k\}$ of S^1 we have

$$(148) \qquad \left\|\sum_k y_k \chi_{T_{e_k}^\delta(a_k)}\right\|_2 \lesssim \sqrt{\log \frac{1}{\delta}}.$$

The relevant geometric fact is

$$(149) \qquad |T_{e_k}^\delta(a) \cap T_{e_l}^\delta(b)| \lesssim \frac{\delta^2}{|e_k - e_l| + \delta}$$

(exercise: prove this). Using (149) we estimate

$$\|\sum_k y_k \chi_{T^\delta_{e_k}(a_k)}\|_2^2 = \sum_{k,l} y_k y_l |T^\delta_{e_k}(a_k) \cap T^\delta_{e_l}(a_l)|$$

$$\lesssim \sum_{k,l} y_k y_l \frac{\delta^2}{|e_k - e_l| + \delta}.$$

$$\lesssim \sum_{k,l} \sqrt{\delta} y_k \sqrt{\delta} y_l \frac{\delta}{|e_k - e_l| + \delta}.$$

(150)

Observe that for fixed k

$$\sum_l \frac{\delta}{|e_l - e_k| + \delta} \lesssim \sum_{l \le \frac{1}{\delta}} \frac{\delta}{l\delta + \delta} = \sum_{l \le \frac{1}{\delta}} \frac{1}{l+1} \approx \log \frac{1}{\delta},$$

and similarly for fixed l

$$\sum_k \frac{\delta}{|e_k - e_l| + \delta} \lesssim \log \frac{1}{\delta}.$$

Applying Schur's test (Lemma 7.5) to the kernel $\delta/(|e_k - e_l| + \delta)$ we obtain that

$$(151) \qquad \|\sum_k y_k \chi_{T^\delta_{e_k}(a_k)}\|_2^2 \lesssim \log \frac{1}{\delta} \sum_k (\sqrt{\delta} y_k)^2 \lesssim \log \frac{1}{\delta},$$

which proves (148). $\qquad\qquad\qquad\qquad\qquad\qquad\qquad\qquad\qquad\qquad$ $\square$

B. Kakeya Problem vs. Restriction Problem

Recall that the restriction conjecture states that

$$\|\widehat{fd\sigma}\|_p \le C_p \|f\|_{L^\infty(S^{n-1})} \text{ if } p > \frac{2n}{n-1}.$$

In fact, the stronger estimate

$$(152) \qquad \|\widehat{fd\sigma}\|_p \le C_p \|f\|_{L^p(S^{n-1})} \text{ if } p > \frac{2n}{n-1}$$

can be proved to be formally equivalent, see e.g. [**3**].

It is known that the restriction conjecture implies the Kakeya conjecture. This is due to Bourgain [**3**], although a related construction had appeared earlier in [**2**]; both constructions are variants on the argument in [**13**].

PROPOSITION 10.5 (Fefferman, Bourgain). *If (152) is true then the conjectured bound*

$$\|f^*_\delta\|_n \le C_\varepsilon \delta^{-\varepsilon} \|f\|_n$$

is also true.

PROOF. We will use Lemma 10.4. Accordingly, we choose a maximal δ-separated set $\{e_j\}$ on S^{n-1}; observe that such a set has cardinality $\approx \delta^{-(n-1)}$. For each j pick a tube $T_{e_j}^{\delta}(a_j)$, and let τ_j be the cylinder obtained by dilating $T_{e_j}^{\delta}(a_j)$ by a factor of δ^{-2} around the origin. Thus τ_j has length δ^{-2}, cross-section radius δ^{-1}, and axis in the e_j direction. Also let

$$S_j = \{e \in S^{n-1} : 1 - e \cdot e_j \le C^{-1}\delta^2\}.$$

Then S_j is a spherical cap of radius approximately $C^{-1}\delta$, centered at e_j. We choose the constant C large enough so that the S_j's are disjoint. Knapp's construction (see Chapter 7) gives a smooth function f_j on S^{n-1} such that f_j is supported on S_j and

$$\|f_j\|_{L^{\infty}(S^{n-1})} = 1,$$

$$|\widehat{f_j d\sigma}| \gtrsim \delta^{n-1} \text{ on } \tau_j.$$

We consider functions of the form

$$f_\omega = \sum_j \omega_j y_j f_j,$$

where y_j are nonnegative coefficients and ω_j are independent random variables taking values ± 1 with equal probability. Since f_j have disjoint supports, we have

$$\|f_\omega\|_{L^q(S^{n-1})}^q = \sum_j \|y_j f_j\|_{L^q(S^{n-1})}^q$$

$$(153) \qquad\qquad \approx \sum_j y_j^q \delta^{n-1}.$$

On the other hand,

$$\mathbb{E}(\|\widehat{f_\omega d\sigma}\|_{L^q(\mathbb{R}^n)}^q) = \int_{\mathbb{R}^n} \mathbb{E}(|\widehat{f_\omega d\sigma}(x)|^q)dx$$

$$\approx \int_{\mathbb{R}^n} (\sum_j y_j^2 |\widehat{f_j d\sigma}(x)|^2)^{q/2} dx \quad \text{(Khinchin's inequality)}$$

$$(154) \qquad\qquad \gtrsim \delta^{q(n-1)} \int_{\mathbb{R}^n} |\sum_j y_j^2 \chi_{\tau_j}(x)|^{q/2} dx.$$

Assume now that (152) is true. Then for any $q > \frac{2n}{n-1}$ it follows from (153) and (154) that

$$\delta^{q(n-1)} \int_{\mathbb{R}^n} |\sum_j y_j^2 \chi_{\tau_j}(x)|^{q/2} dx \lesssim \sum_j y_j^q \delta^{n-1}.$$

Let $z_j = y_j^2$ and $p' = q/2$, then the above inequality is equivalent to the statement

$$(155) \qquad \text{if } \delta^{n-1} \sum_j z_j^{q/2} \le 1, \text{ then } \|\sum_j z_j \chi_{\tau_j}\|_{q/2} \lesssim \delta^{-2(n-1)}$$

for any $p' \geq \frac{n}{n-1}$. We now rescale this by δ^2 to obtain

$$\text{if } \delta^{n-1} \sum_j z_j^{p'} \leq 1, \text{ then } \|\sum_j z_j \chi_{T_j}\|_{p'} \lesssim \delta^{2(\frac{n}{p'}-(n-1))}.$$

Observe that $\frac{n}{p'} - (n-1) \downarrow 0$ as $p' \downarrow \frac{n}{n-1}$. Thus for any $\varepsilon > 0$ we have

$$(156) \qquad \text{if } \delta^{n-1} \sum_j z_j^{p'} \leq 1, \text{ then } \|\sum_j z_j \chi_{T_j}\|_{p'} \lesssim \delta^{-\varepsilon}$$

if p' is close enough to $\frac{n}{n-1}$. By Lemma 10.4, this implies that for any $\varepsilon > 0$

$$\|f_\delta^*\|_p \lesssim \delta^{-\varepsilon} \|f\|_p$$

provided that $p < n$ is close enough to n. Interpolating this with the trivial L^∞ bound, we conclude that

$$\|f_\delta^*\|_n \lesssim \delta^{-\varepsilon} \|f\|_n$$

as claimed. $\qquad\square$

We proved that the restriction conjecture is stronger than the Kakeya conjecture. Bourgain [3] partially reversed this and obtained a restriction theorem beyond Stein-Tomas by using a Kakeya set estimate that is stronger than the L^2 bound (151) used in the proof of (147). It is not known whether (either version of) the Kakeya conjecture implies the full restriction conjecture.

THEOREM 10.6 (Bourgain [3]). *Suppose that we have an estimate*

$$(157) \qquad \|\sum_j \chi_{T_{e_j}^\delta(a_j)}\|_{q'} \leq C_\varepsilon \delta^{-(\frac{n}{q}-1+\varepsilon)}$$

for any given $\varepsilon > 0$ and for some fixed $q > 2$. Then

$$(158) \qquad \|\widehat{fd\sigma}\|_p \leq C_p \|f\|_{L^\infty(S^{n-1})}$$

for some $p < \frac{2n+2}{n-1}$.

REMARK The geometrical statement corresponding to (157) is that Kakeya sets in $\mathbb{R}^n$ have dimension at least q.

We will sketch the proof only for $n = 3$. Recall that in $\mathbb{R}^3$ we have the estimates

$$(159) \qquad \|\widehat{fd\sigma}\|_4 \lesssim \|f\|_{L^2(S^2)}$$

from the Stein-Tomas theorem, and

$$(160) \qquad \|\widehat{fd\sigma}\|_{L^2(D(0,R))} \lesssim R^{1/2} \|f\|_{L^2(S^2)}$$

from Theorem 7.4 with $\alpha = n - 1 = 2$. Interpolating (159) and (160) yields a family of estimates

$$(161) \qquad \|\widehat{fd\sigma}\|_{L^p(D(0,R))} \lesssim R^{\frac{2}{p}-\frac{1}{2}} \|f\|_{L^2(S^2)}$$

for $2 \leq p \leq 4$. Below we sketch an argument showing that the exponent of R in (161) can be lowered if the L^2 norm on the right side is replaced by the L^∞ norm.

PROPOSITION 10.7. *Let $n = 3$, $2 < p < 4$, and assume that (157) holds for some $q > 2$. Then*

$$\|\widehat{fd\sigma}\|_{L^p(D(0,R))} \lesssim R^{\alpha(p)}\|f\|_{L^\infty(S^{n-1})},$$

where $\alpha(p) < \frac{2}{p} - \frac{1}{2}$.

This of course implies (158) for all p such that $\alpha(p) \leq 0$; in particular, there are $p < 4$ for which (158) holds.

HEURISTIC PROOF OF THE PROPOSITION Assume that $\|f\|_{L^\infty(S^{n-1})} = 1$, and let $\delta = R^{-1}$. We cover S^2 by spherical caps

$$S_j = \{e \in S^2 : 1 - e \cdot e_j \leq \delta\},$$

where $\{e_j\}$ is a maximal $\sqrt{\delta}$-separated set on S^2. Then

$$f = \sum_j f_j,$$

where each f_j is supported on S_j. Let $G = \widehat{fd\sigma}$ and $G_j = \widehat{f_j d\sigma}$, so that $G = \sum_j G_j$. By the uncertainty principle $|G_j|$ is roughly constant on cylinders of length R and diameter $\sqrt{R}$ pointing in the e_j direction. To simplify the presentation, we now make the assumption[1] that G_j is supported on only one such cylinder τ_j.

Next, we cover $D(0, R)$ with disjoint cubes Q of side $\sqrt{R}$. For each Q we denote by $N(Q)$ the number of cylinders τ_j which intersect it. Note that $G|_Q = \sum_j G_j|_Q$, where we sum only over those j's for which τ_j intersects Q. Using this and (161), we can estimate $\|G\|_{L^p(Q)}$ for $2 \leq p \leq 4$:

$$\|G\|_{L^p(Q)} \lesssim \sqrt{R}^{\frac{2}{p}-\frac{1}{2}}\left\|\sum_{j:\tau_j \cap Q \neq \emptyset} f_j\right\|_{L^2(S^2)}$$

$$\lesssim \sqrt{R}^{\frac{2}{p}-\frac{1}{2}}(N(Q) \cdot |S_i|)^{1/2}$$

$$\approx \delta^{\frac{3}{4}-\frac{1}{p}}N(Q)^{1/2}. \tag{162}$$

Summing over Q, we obtain

$$\|G\|_{L^p(D(0,R))}^p \lesssim \delta^{\frac{3p}{4}-1}\sum_Q N(Q)^{p/2}$$

$$\approx \delta^{\frac{3p}{4}p+\frac{1}{2}}\left\|\sum_j \chi_{\tau_j}\right\|_{p/2}^{p/2}. \tag{163}$$

---

[1]It is because of this assumption that our proof is merely heuristic. Of course the Fourier transform of a compactly supported measure cannot be compactly supported; the rigorous proof uses the Schwartz decay of G_j instead.

On the last line we used that

$$\|\sum_j \chi_{\tau_j}\|_{p/2}^{p/2} = \sum_Q N(Q)^{p/2} \cdot |Q| = \delta^{-3/2} \sum_Q N(Q)^{p/2}.$$

We now let $p = 2q'$, where q' is the exponent in (157), and assume that p is sufficiently close to 4 (interpolate (157) with (147) if necessary). We have from (157)

$$\|\sum_j \chi_{T_{e_j}^{\sqrt{\delta}}(a_j)}\|_{q'} \le C_\varepsilon \sqrt{\delta}^{-\left(\frac{3}{q}-1+\varepsilon\right)}.$$

Rescaling this inequality by δ^{-1} we obtain

$$\|\sum_j \chi_{\tau_j}\|_{q'} \lesssim \sqrt{\delta}^{-\left(\frac{3}{q}-1+\varepsilon\right)} \delta^{-3/q'} = \delta^{-1-\frac{3}{p}-\varepsilon}.$$

We combine this with (162) and conclude that

$$\|G\|_{L^p(Q)}^p \lesssim \delta^{\frac{p}{4}-1}\delta^{-\varepsilon},$$

i.e.,

$$\|\widehat{fd\sigma}\|_{L^p(D(0,R))} \lesssim \delta^{\frac{1}{4}-\frac{1}{p}-\varepsilon} = R^{\frac{1}{p}-\frac{1}{4}+\varepsilon},$$

which proves the proposition since $\frac{1}{p} - \frac{1}{4} < \frac{2}{p} - \frac{1}{2}$ if $p < 4$. $\qquad\square$

Bibliography

[1] W. Beckner, *Inequalities in Fourier analysis*, Ann. of Math. 102 (1975), 159-182.

[2] W. Beckner, A. Carbery, S. Semmes, F. Soria: *A note on restriction of the Fourier transform to spheres*, Bull. London Math. Soc. 21 (1989), 394-398.

[3] J. Bourgain, *Besicovitch type maximal operators and applications to Fourier analysis*, Geom. Funct. Anal. 1 (1991), 147-187.

[4] J. Bourgain, *Hausdorff dimension and distance sets*, Israel J. Math 87 (1994), 193-201.

[5] J. Bourgain, L^p *estimates for oscillatory integrals in several variables*, Geom. Funct. Anal. 1 (1991), 321-374.

[6] L. Carleson, *Selected Problems on Exceptional Sets*, Van Nostrand Mathematical Studies. No. 13, Van Nostrand Co., Inc., Princeton-Toronto-London 1967.

[7] A. Córdoba, *The Kakeya maximal function and spherical summation multipliers*, Amer. J. Math. 99 (1977), 1-22.

[8] R. O. Davies, *Some remarks on the Kakeya problem*, Proc. Cambridge Phil. Soc. 69(1971), 417-421.

[9] K.M. Davis, Y.C. Chang, *Lectures on Bochner-Riesz Means* London Mathematical Society Lecture Note Series, 114, Cambridge University Press, Cambridge, 1987.

[10] K. J. Falconer, *The geometry of fractal sets*, Cambridge University Press, 1985.

[11] K. J. Falconer, *On the Hausdorff dimension of distance sets*, Mathematika 32(1985), 206-212.

[12] C. Fefferman, *Inequalities for strongly singular convolution operators*, Acta Math. 124(1970), 9-36.

[13] C. Fefferman, *The multiplier problem for the ball*, Ann. of Math. 94(1971), 330-336.

[14] G. Folland, *Harmonic Analysis in Phase Space*, Annals of Mathematics Studies, 122. Princeton University Press, Princeton, NJ, 1989.

[15] V. Havin, B. Joricke, *The Uncertainty Principle in Harmonic Analysis*, Springer-Verlag, 1994.

[16] L. Hörmander, *Oscillatory integrals and multipliers on FL^p*, Ark. Mat. 11 (1973), 1-11.

[17] L. Hörmander, *The Analysis of Linear Partial Differential Operators*, volume 1, 2nd edition, Springer Verlag 1990.

[18] N. Katz, I. Laba, T. Tao, *An improved bound on the Minkowski dimension of Besicovitch sets in $\mathbb{R}^3$*, Ann. of Math. 152 (2000), 383-446.

[19] N. Katz, T. Tao, *New bounds for Kakeya sets*, J. Anal. Math. 87 (2002), 231–263.

[20] Y. Katznelson, *An Introduction to Harmonic Analysis* 2nd edition. Dover Publications, Inc., New York, 1976.

[21] R. Kaufman, *On the theorem of Jarnik and Besicovitch*, Acta Arithmetica 39 (1981), 265-267.

[22] I. Laba, T. Tao, *An improved bound for the Minkowski dimension of Besicovitch sets in medium dimension*, Geom. Funct. Anal. 11(2001), 773-806.

[23] J. E. Marsden, *Elementary Classical Analysis*, W. H. Freeman and Co., San Francisco, 1974

[24] P. Mattila, *Spherical averages of Fourier transforms of measures with finite energy; dimension of intersections and distance sets*, Mathematika 34 (1987), 207-228.

[25] P. Mattila, *Geometry of sets and measures in Euclidean spaces*, Cambridge University Press, 1995.

[26] W. Minicozzi, C. Sogge, *Negative results for Nikodym maximal functions and related oscillatory integrals in curved space*, Math. Res. Lett. 4 (1997), 221–237.

[27] W. Rudin, *Functional Analysis*, McGraw-Hill, Inc., New York, 1991.

[28] R. Salem, *Algebraic Numbers and Fourier Analysis*, D.C. Heath and Co., Boston, Mass. 1963.

[29] C. Sogge, *Fourier integrals in Classical Analysis*, Cambridge University Press, Cambridge, 1993.

[30] C. Sogge, *Concerning Nikodym-type sets in 3-dimensional curved space*, J. Amer. Math. Soc., 1999.

[31] J. Solymosi, C. Tóth, *Distinct distances in the plane*, Discrete Comput. Geometry 25 (2001), 629-634.

[32] E.M. Stein, in *Beijing Lectures in Harmonic Analysis*, edited by E. M. Stein.

[33] E. M. Stein, *Harmonic Analysis*, Princeton University Press 1993.

[34] E. M. Stein, G. Weiss, *Introduction to Fourier Analysis on Euclidean Spaces*, Princeton University Press, Princeton, N.J. 1990

[35] P. A. Tomas, *A restriction theorem for the Fourier transform*, Bull. Amer. Math. Soc. 81 (1975), 477-478.

[36] A. Varchenko, *Newton polyhedra and estimations of oscillatory integrals*, Funct. Anal. Appl. 18 (1976), 175-196.

[37] T. Wolff, *Decay of circular means of Fourier transforms of measures*, Internat. Math. Res. Notices 10 (1999), 547-567.

[38] T. Wolff, *An improved bound for Kakeya type maximal functions*, Rev. Mat. Iberoamericana 11 (1995), 651–674.

[39] T. Wolff, *Recent work connected with the Kakeya problem*, in *Prospects in Mathematics*, H. Rossi, ed., Amer. Math. Soc., Providence, R.I. (1999), 129-162.

[40] A. Zygmund, *Trigonometric Series*, Cambridge University Press, Cambridge, 1968.

CHAPTER 11

Recent Work Connected with the Kakeya Problem[1]

A *Kakeya set* in $\mathbb{R}^n$ is a compact set $E \subset \mathbb{R}^n$ containing a unit line segment in every direction, i.e.

$$(164) \qquad \forall e \in S^{n-1} \ \exists x \in \mathbb{R}^n : x + te \in E \ \ \forall t \in [-\tfrac{1}{2}, \tfrac{1}{2}]$$

where S^{n-1} is the unit sphere in $\mathbb{R}^n$. This paper will be mainly concerned with the following issue, which is still poorly understood: what metric restrictions does the property (164) put on the set E?

The original Kakeya problem was essentially whether a Kakeya set as defined above must have positive measure, and as is well-known, a counterexample was given by Besicovitch in 1920. A current form of the problem is as follows:

OPEN QUESTION 1. Must a Kakeya set in $\mathbb{R}^n$ have Hausdorff dimension n?

When $n = 2$, the answer is yes; this was proved by Davies [21] in 1971. Recent work on the higher dimensional question began with [7]. If $\dim E$ denotes the Hausdorff dimension then the bound $\dim E \geq \frac{n+1}{2}$ can be proved in several ways and may have been known prior to [7], although the author has not been able to find a reference. The recent work [7], [73] has led to the small improvement $\dim E \geq \frac{n+2}{2}$. We will discuss this in Section 2 below[2].

Question 1 appears quite elementary, but is known to be connected to a number of basic open questions in harmonic analysis regarding estimation of oscillatory integrals. This is a consequence of C. Fefferman's solution of the disc multiplier problem [25] and the work of Córdoba (e.g. [20]) and Bourgain (e.g. [7], [9], [10]). We will say something about these interrelationships in Section 4. There is also a long history of applications of Kakeya sets to construct counterexamples in pointwise convergence questions; we will not discuss this here, but see e.g. [27] and [64].

For various reasons it is better to look also at a more quantitative formulation in terms of a maximal operator. If $\delta > 0$, $e \in S^{n-1}$, $a \in \mathbb{R}^n$ then

---

[1]Reprinted from *Prospects in Mathematics: Invited Talks on the Occasion of the 250th Anniversary of Princeton University*, Hugo Rossi (editor), American Mathematical Society, 1999.

[2]*Editor's note:* For further references see the footnote to Remark 2.2.

91

we define

$$T_e^\delta(a) = \{x \in \mathbb{R}^n : |(x-a)\cdot e| \le \tfrac{1}{2}, \ |(x-a)^\perp| \le \delta\}$$

where $x^\perp = x - (x\cdot e)e$. Thus $T_e^\delta(a)$ is essentially the δ-neighborhood of the unit line segment in the e direction centered at a. If $f : \mathbb{R}^n \to \mathbb{R}$ then we define its Kakeya maximal function $f_\delta^* : S^{n-1} \to \mathbb{R}$ via

$$f_\delta^*(e) = \sup_{a\in\mathbb{R}^n} \frac{1}{|T_e^\delta(a)|} \int_{T_e^\delta(a)} |f|$$

This definition is due to Bourgain [7]. It is one of several similar maximal functions that have been considered, going back at least to [20].

OPEN QUESTION 2. Is there an estimate

$$(165) \qquad \forall \epsilon > 0 \ \exists C_\epsilon : \|f_\delta^*\|_{L^n(S^{n-1})} \le C_\epsilon \delta^{-\epsilon} \|f\|_n \ \forall f$$

Roughly speaking, this question is related to Question 1 in the same way as the Hardy–Littlewood maximal theorem is related to Lebesgue's theorem on points of density. As was observed by Bourgain [7], an affirmative answer to Question 2 implies an affirmative answer to Question 1; see Lemma 11.9 below. Once again, when $n = 2$ it is well known that the answer to Question 2 is affirmative, [20] and [7]. In higher dimensions, partial results are known paralleling the results on Question 1.

Questions 1 and 2 clearly have a combinatorial side to them, and the point of view we will adopt here is to try to approach the combinatorial issues directly using ideas from the combinatorics literature. In this connection let us mention a basic principle in graph theory (the "Zarankiewicz problem"; see [5], [26], [50] for this and generalizations): fix s and suppose that $\{a_{ij}\}_{i=1}^n{}_{j=1}^m$ is an $n \times m$ $(0,1)$ matrix with no $s \times s$ submatrix of 1's. Then there is a bound

$$(166) \qquad |\{(i,j) : a_{ij} = 1\}| \le C_s \min(mn^{1-\frac{1}{s}} + n, nm^{1-\frac{1}{s}} + m)$$

To see the relationship between this sort of bound and Kakeya, just note that if $\{\ell_j\}_{j=1}^m$ are lines and $\{p_i\}_{i=1}^n$ are points, then the "incidence matrix"

$$a_{ij} = \begin{cases} 1 & \text{if } p_i \in \ell_j \\ 0 & \text{if } p_i \notin \ell_j \end{cases}$$

will contain no 2×2 submatrix of 1's, since two lines intersect in at most one point. Much of what we will say below will have to do with attempts to modify this argument, and also more sophisticated arguments in incidence geometry (e.g. [19]) to make them applicable to "continuum" problems such as Kakeya.

There are several difficulties with such an approach. It is sometimes unclear whether applying the combinatorial techniques in the continuum should be simply a matter of extra technicalities or whether new phenomena should be expected to occur, and furthermore many of the related discrete problems are regarded as being very difficult. A classical example is the

Erdős unit distance problem (see [19] and [50]) and other examples will be mentioned in Section 3.

Of course, much work has been done in the opposite direction, applying harmonic analysis techniques to questions of a purely geometrical appearance. A basic example is the spherical maximal theorem of Stein [61], and various Strichartz type inequalities as well as the results on the distance set problem in [24], [11] are also fairly close to the subject matter of this paper. However, we will not present any work of this nature here.

The paper is organized as follows. In Section 1 we discuss the two dimensional Kakeya problem, in Section 2 we discuss the higher dimensional Kakeya problem and in Section 3 we discuss analogous problems for circles in the plane. Finally in Section 4 we discuss the Fefferman construction and a related construction of Bourgain [9] which connects the Kakeya problem also to estimates of Dirichlet series. Section 4 contains several references to the recent literature on open problems regarding oscillatory integrals, but it is not a survey. Further references are in [10], [68], and especially [62].

We have attempted to make the presentation self-contained insofar as is possible. In particular we will present some results and arguments which are known or almost known but for which there is no easy reference.

The author is grateful for the opportunity to speak at the conference and to publish this article.

List of notation

$[\alpha]$: greatest integer less than or equal to α.

p' : conjugate exponent to p, i.e. $p' = \frac{p}{p-1}$.

$D(x,r)$: the disc with center x and radius r.

$|E|$: Lebesgue measure or cardinality of the set E, depending on the context.

E^c : complement of E.

$\dim E$: Hausdorff dimension of E.

$H_s(E)$: s-dimensional Hausdorff content of E, i.e. $H_s(E) = \inf\left(\sum_j r_j^s : E \subset \bigcup_j D(x_j, r_j)\right)$.

$T_e^\delta(a)$: δ-tube in the e direction centered at a, as defined in the introduction. Sometimes we will also use the notation T_e^δ; this means any tube of the form $T_e^\delta(a)$ for some $a \in \mathbb{R}^n$.

$C(x,r)$: circle in $\mathbb{R}^2$ (or sphere in $\mathbb{R}^n$) with center x and radius r.

$C_\delta(x,r)$: annular region $\{y \in \mathbb{R}^n : r - \delta < |y - x| < r + \delta\}$.

$x \lesssim y$: $x \leq Cy$ for a suitable constant C.

11.1. The two dimensional case

We will start by proving the existence of measure zero Kakeya sets using a variant on the original construction which is quick and is easy to write out in closed form; to the author's knowledge the earliest reference for this approach is Sawyer [**52**]. A discussion of various other possible approaches to the construction may be found in [**23**].

For expository reasons, we make the following definitions.

A *G-set* is a compact set $E \subset \mathbb{R}^2$ which is contained in the strip $\{(x, y) : 0 \le x \le 1\}$, such that for any $m \in [0, 1]$ there is a line segment contained in E connecting $x = 0$ to $x = 1$ with slope m, i.e.

$$\forall m \in [0, 1] \; \exists b \in \mathbb{R} : mx + b \in E \; \forall x \in [0, 1].$$

If $\ell = \{(x, y) : y = mx + b\}$ is a nonvertical line and $\delta > 0$, then $S_\ell^\delta \stackrel{\text{def}}{=} \{(x, y) : 0 \le x \le 1 \text{ and } |y - (mx + b)| \le \delta\}$.

REMARK 11.1. It is clear that existence of G-sets with measure zero will imply existence of Kakeya sets with measure zero. Note also that if ℓ is a line with slope m then S_ℓ^δ will contain segments connecting $x = 0$ to $x = 1$ with any given slope between $m - 2\delta$ and $m + 2\delta$.

We now describe the basic construction, which leads to the slightly weaker conclusion (Lemma 11.3) that there are G-sets with measure $< \epsilon$ for any $\epsilon > 0$. It can be understood in terms of the usual sliding triangle picture: start from a right triangle with vertices $(0, 0)$, $(0, -1)$ and $(1, 0)$; this is clearly a *G-set*. Subdivide it in N "1st stage" triangles by subdividing the vertical side in N equal intervals. Leave the top triangle alone and slide the others upward so that their intersections with the line $x = 0$ all coincide. Next, for each of the 1st stage triangles, subdivide it in N 2nd stage triangles, leave the top triangle in each group alone and slide the $N - 1$ others upward until the intersections of the N triangles in the group with the line $x = \frac{1}{N}$ all coincide. Now repeat at abscissas $\frac{2}{N}$, $\frac{3}{N}$, ..., $\frac{N-1}{N}$.

Now we make this precise. Fix a large integer N and let $\mathcal{A}_N$ be all numbers in $[0, 1)$ whose base N expansion terminates after N digits, i.e.

$$a \in \mathcal{A}_N \Leftrightarrow a = \sum_{j=1}^{N} \frac{a_j}{N^j} \text{ with } a_j \in \{0, 1, \ldots N - 1\}.$$

To each $a \in \mathcal{A}_N$ we associate the line segment ℓ_a connecting the y axis to the line $x = 1$ with slope a and y intercept $-\sum_{j=1}^{N} \frac{(j-1)a_j}{N^{j+1}}$. Thus

$$\ell_a = \{(t, \phi_a(t)) : 0 \le t \le 1\}, \text{ where } \phi_a(t) = \sum_{j=1}^{N} \frac{(Nt - j + 1)a_j}{N^{j+1}}.$$

LEMMA 11.2. *For each $t \in [0, 1]$ there are an integer $k \in \{1, \ldots, N\}$ and a set of N^{k-1} intervals each of length $2N^{-k}$, whose union contains the set $\{\phi_a(t) : a \in \mathcal{A}_N\}$.*

PROOF. Choose k so that $\frac{k-1}{N} \le t \le \frac{k}{N}$. Define $a, b \in \mathcal{A}_N$ to be equivalent if $a_j = b_j$ when $j \le k - 1$. There are N^{k-1} equivalence classes, and if a

and b are equivalent then

$$|\phi_a(t) - \phi_b(t)| = \left| \sum_{j \geq k} \frac{(Nt - j + 1)(a_j - b_j)}{N^{j+1}} \right| \leq \sum_{j \geq k} \frac{\max(j - k, 1)|a_j - b_j|}{N^{j+1}}$$

$$\leq \frac{N - 1}{N^{k+1}} \sum_{j \geq k} \frac{\max(j - k, 1)}{N^{j-k}} \leq 2 \frac{N - 1}{N^{k+1}}$$

$$< \frac{2}{N^k} \quad \text{when } N \text{ is large.}$$

$\square$

LEMMA 11.3. *Let N be sufficiently large. Then there is a G-set $E \subset [0,1] \times [-1,1]$ which intersects every vertical line in measure $\leq \frac{4}{N}$, in particular $|E| \leq \frac{4}{N}$.*

PROOF. We let

$$E_N = \bigcup_{a \in \mathcal{A}_N} S_{\ell_a}^{N^{-N}}.$$

Then E_N contains segments with all slopes between 0 and 1, by Remark 11.1. If $t \in [0,1]$, then by Lemma 11.2 there is $k \in \{1, \ldots, N\}$ such that the intersection of E with the line $x = t$ is contained in the union of N^{k-1} intervals of length $2N^{-k} + 2N^{-N} \leq 4N^{-k}$. The lemma follows. $\square$

Existence of measure zero Kakeya sets now follows by a standard limiting argument, most easily carried out via the following lemma.

LEMMA 11.4. *For every G-set E and every $\epsilon > 0, \eta > 0$, there is another G-set F, which is contained in the ϵ-neighborhood of E and has measure $< \eta$.*

PROOF. Let δ be small, let $\{m_j\} = \{j\delta\}_{j=0}^{[\frac{1}{\delta}]}$ and for each j, fix a line segment $\ell_j = \{(x,y) : 0 \leq x \leq 1, y = m_j x + b_j\} \subset E$ with slope m_j connecting $x = 0$ to $x = 1$ and form the parallelogram $S_{\ell_j}^\delta$. Let A_j be the affine map from $[0,1] \times [-1,1]$ on $S_{\ell_j}^\delta$, $A_j(x,y) = (x, m_j x + b_j + \delta y)$ and consider $F = \bigcup_m A_m(E_N)$ for a large enough N; here E_N is as in Lemma 11.3. A_j maps segments with slope μ to segments with slope $m + \delta\mu$ so F is a G-set. Clearly it is contained in the δ-neighborhood of E. Furthermore A_j contracts areas by a factor δ so $|A_j(E_N)| \leq 4\frac{\delta}{N}$ for each j, hence $|F| \lesssim \frac{1}{N}$. $\square$

COROLLARY 11.5. *There are Kakeya sets with measure zero.*

PROOF. We construct a sequence $\{F_n\}_{n=0}^\infty$ of G-sets, and a sequence of numbers $\{\epsilon_n\}_{n=0}^\infty$ converging to zero such that the following properties hold when $n \geq 1$; here $F(\epsilon) \stackrel{\text{def}}{=} \{x : \text{dist}(x, F) < \epsilon\}$ is the ϵ-neighborhood of F and $\overline{E}$ is the closure of E.

(1) $F_n(\epsilon_n) \subset F_{n-1}(\epsilon_{n-1})$.
(2) $|\overline{F_n(\epsilon_n)}| < 2^{-n}$.

Namely, we take F_0 to be any G-set, and we set $\epsilon_0 = 1$. If $n \geq 1$ and if F_{n-1} and ϵ_{n-1} have been constructed then we obtain F_n by applying Lemma 11.4 with $\epsilon = \epsilon_{n-1}$ and $\eta = 2^{-n}$. Since F_n is compact, (i) and (ii) will then hold provided ϵ_n is sufficiently small.

The set $\bigcap_n \overline{F_n(\epsilon_n)}$ is then a G-set with measure zero. $\square$

REMARK 11.6. The construction above easily gives the following variant (used e.g. in [25]): with $\delta = \frac{1}{10} N^{-N}$, there is a family of disjoint δ-tubes

$$\{T_{e_j}^\delta(x_j)\}_{j=1}^M \subset \mathbb{R}^2$$

where $M \approx \delta^{-1}$ with the property that the union of the translated tubes $T_{e_j}^\delta(x_j + 2e_j)$ has measure $\lesssim \frac{1}{N}$.

Namely, a calculation shows that if $a, b \in \mathcal{A}$ and $a < b$ then $\phi_a(1) < \phi_b(1)$, i.e. the ordering of the intersection points between the ℓ_a and the line $x = 1$ is the same as the ordering of slopes. Hence if we regard ℓ_a as extended to a complete line, then no two ℓ_a's intersect in the region $x > 1$, and in fact in the region $x > 2$ any two of them are at least N^{-N} apart. Now for each $a \in \mathcal{A}_N$ we form the rectangle R_a with length 1, width $\frac{1}{5} N^{-N}$, axis along the line ℓ_a and bottom right corner on the line $x = 1$. Clearly $R_a \subset S_a$, so $\bigcup_a R_a$ is small by Lemma 11.3. On the other hand, if R_a is translated to the right along its axis by distance 2 then the resulting rectangles are disjoint. We may therefore take $\{T_{e_j}^\delta(x_j)\}$ to be the set of translated rectangles.

REMARK 11.7. Analogous statements in higher dimensions may be obtained using dummy variables.

Measure zero Kakeya sets in $\mathbb{R}^n$ may be constructed by taking the product of a Kakeya set in $\mathbb{R}^2$ with a closed disc of radius $\frac{1}{2}$ in $\mathbb{R}^{n-2}$ (or for that matter with any Kakeya set in $\mathbb{R}^{n-2}$), and a family of roughly $\delta^{-(n-1)}$ disjoint $T_e^\delta(a)$'s such that the union of the tubes $T_e^\delta(a + 2e)$ has small measure may be obtained by taking the products of the tubes in Remark 11.6 with a family of $\delta^{-(n-2)}$ disjoint δ-discs in $\mathbb{R}^{n-2}$.

We now discuss the positive results on Questions 1 and 2 in dimension two. Proposition 11.8 was first stated and proved in [7] although a similar result for a related maximal function was proved earlier in [20].

We will work with restricted weak type estimates instead of with L^p estimates; this is known to be equivalent except for the form of the $\delta^{-\epsilon}$ terms.[3] We will say (see e.g. [63]) that an operator T has restricted weak type norm $\leq A$, written

$$\|Tf\|_{q,\infty} \leq A\|f\|_{p,1}$$

---

[3] We work with restricted weak type estimates for expository reasons only. We believe this makes the results more transparent; however, it is well known that actually $\|f_\delta^*\|_{L^2(S^1)} \lesssim (\log\frac{1}{\delta})^{\frac{1}{2}}\|f\|_2$. The latter estimate is proved in [7] and also follows from the proof below, plus duality, as in [20].

if $\left|\{x : |T\chi_E(x)| \geq \lambda\}\right| \leq \left(\frac{A|E|^{\frac{1}{p}}}{\lambda}\right)^q$ for all sets E with finite measure and all $\lambda \in (0,1]$; here χ_E is the characteristic function of E.

PROPOSITION 11.8. *The restricted weak type $(2,2)$ norm of the Kakeya maximal operator $f \to f_\delta^*$ in $\mathbb{R}^2$ is $\lesssim (\log \frac{1}{\delta})^{\frac{1}{2}}$.*

More explicitly, suppose that $E \subset \mathbb{R}^2$ and $\lambda \in (0,1]$. Let $f = \chi_E$, and let $\Omega = \{e \in S^1 : f_\delta^*(e) \geq \lambda\}$. Then

$$|\Omega| \lesssim \log \frac{1}{\delta} \frac{|E|}{\lambda^2}.$$

PROOF. Let $\theta(e,f)$ be the unoriented angle subtended by the directions e and f, i.e. $\theta(e,f) = \arccos(e \cdot f)$. We start by mentioning two trivial but important facts. First, in $\mathbb{R}^n$, the intersection of the tubes $T_e^\delta(a)$ and $T_f^\delta(b)$ satisfies

$$(167) \qquad \operatorname{diam}(T_e^\delta(a) \cap T_f^\delta(b)) \lesssim \frac{\delta}{\theta(e,f) + \delta}$$

for any a and b and therefore also

$$(168) \qquad |T_e^\delta(a) \cap T_f^\delta(b)| \lesssim \frac{\delta^n}{\theta(e,f) + \delta}.$$

Next, if Ω is a set on the unit sphere $S^{n-1} \subset \mathbb{R}^n$ and if $\delta > 0$ then the δ-entropy $\mathcal{N}_\delta(\Omega)$ (maximum possible cardinality M for a δ-separated subset $\{e_j\}_{j=1}^M \subset \Omega$) satisfies

$$(169) \qquad \mathcal{N}_\delta(\Omega) \gtrsim \frac{|\Omega|}{\delta^{n-1}}.$$

Now we assume $n = 2$ and give the proof of the proposition. Fix a δ-separated $\{e_j\}_{j=1}^M \subset \Omega$, where $M \gtrsim \frac{|\Omega|}{\delta}$. For each j, there is a tube $T_j = T_{e_j}^\delta(a_j)$ with $|T_j \cap E| \geq \lambda |T_j| \approx \lambda \delta$. Thus

$$M\lambda\delta \lesssim \sum_j |T_j \cap E| = \int_E \sum_j \chi_{T_j} \leq |E|^{\frac{1}{2}} \left\|\sum_j \chi_{T_j}\right\|_2$$

$$= |E|^{\frac{1}{2}} \left(\sum_{j,k} |T_j \cap T_k|\right)^{\frac{1}{2}} \lesssim |E|^{\frac{1}{2}} \left(\sum_{j,k} \frac{\delta^2}{\theta(e_j, e_k) + \delta}\right)^{\frac{1}{2}}.$$

For fixed k the sum over j is $\lesssim \sum_{j:|j-k|\leq\frac{C}{\delta}} \frac{\delta^2}{|j-k|\delta+\delta} \lesssim \delta \log \frac{1}{\delta}$. We conclude that $M\lambda\delta \lesssim |E|^{\frac{1}{2}} (M\delta \log \frac{1}{\delta})^{\frac{1}{2}}$ which gives the result since $M \gtrsim \frac{|\Omega|}{\delta}$. $\square$

Now we show how to pass to the Hausdorff dimension statement. The next result is Lemma 2.15 in [**7**].

LEMMA 11.9. *Assume an estimate in $\mathbb{R}^n$*

$$(170) \qquad \|f_\delta^*\|_{q,\infty} \leq C\delta^{-\alpha}\|f\|_{p,1}.$$

Then Kakeya sets have dimension at least $n - p\alpha$.

PROOF. Fix $s < n - p\alpha$. Let E be a Kakeya set and for each $e \in S^{n-1}$, fix a point x_e such that $x_e + te \in E$ when $t \in [-\frac{1}{2}, \frac{1}{2}]$. We have to bound $H_s(E)$ from below, so fix a covering of E by discs $D_j = D(x_j, r_j)$. We can evidently suppose all r_j's are ≤ 1.

Let $\Sigma_k = \{j : 2^{-k} \leq r_j \leq 2^{-(k-1)}\}$, $\nu_k = |\Sigma_k|$ and $E_k = E \cap (\bigcup\{D_j : j \in \Sigma_k\})$. Also let $\tilde{D}_j = D(x_j, 2r_j)$, and $\tilde{E}_k = \bigcup\{\tilde{D}_j : j \in \Sigma_k\}$.

Then $\bigcup_k E_k = E$, so for each e the pigeonhole principle implies

$$\left|\{t \in [-\tfrac{1}{2}, \tfrac{1}{2}] : x_e + te \in E_k\}\right| \geq \frac{c}{k^2}$$

for some $k = k_e$, where $c = \frac{6}{\pi^2}$. By the pigeonhole principle again, we can find a fixed k so that $k = k_e$ when $e \in \Omega$, where $\Omega \subset S^{n-1}$ has measure $\geq \frac{c}{k^2}$. With this k, we note that $\tilde{E}_k$ contains a disc of radius 2^{-k} centered at each point of E_k; it follows easily that if $e \in \Omega$ then $\left|T_e^{2^{-k}}(x_e) \cap \tilde{E}_k\right| \gtrsim k^{-2}\left|T_e^{2^{-k}}(x_e)\right|$. With $f = \chi_{\tilde{E}_k}$ we therefore have

$$\left|\{e : f_{2^{-k}}^*(e) \geq C^{-1}k^{-2}\}\right| \gtrsim k^{-2}.$$

On the other hand, by the assumption (170)

$$\left|\{e : f_{2^{-k}}^*(e) \geq C^{-1}k^{-2}\}\right| \lesssim \left(k^2 2^{k\alpha}|\tilde{E}_k|^{\frac{1}{p}}\right)^q$$

and $|\tilde{E}_k| \lesssim \nu_k 2^{-kn}$. So $\left(k^2 2^{k\alpha}(\nu_k 2^{-kn})^{\frac{1}{p}}\right)^q \gtrsim k^{-2}$, or equivalently

$$\nu_k \gtrsim k^{-\frac{2}{p}(1+\frac{1}{q})} 2^{k(n-p\alpha)}.$$

Letting $\epsilon = n - p\alpha - s > 0$, we have $\sum_j r_j^s \gtrsim \nu_k 2^{-ks} \gtrsim k^{-2p(1+\frac{1}{q})} 2^{k\epsilon} \geq$ constant. $\qquad\qquad\square$

Applying this with $p = n = 2$ we see that Proposition 11.8 implies Davies' theorem that Kakeya sets in $\mathbb{R}^2$ have dimension 2. Likewise it follows that yes on Question 2 for a given n will imply yes on Question 1 for the same n.

REMARK 11.10. It is clear that the logarithmic factor in Proposition 11.8 cannot be dropped entirely, since then the above argument would show that measure zero Kakeya sets could not exist. In fact it has been known for a long time that the exponent $\frac{1}{2}$ cannot be improved, and U. Keich [34] recently showed that even a higher order improvement is not possible in Proposition 11.8 or in its corollary on L^p for $p > 2$. On the other hand, a number of related questions concerning logarithmic factors have been solved only recently or are still open. In particular we should mention the results of Barrionuevo [2] and Katz [28], [29], [30] on the question of maximal functions defined using families of directions in the plane.

REMARK 11.11. An interesting open question in $\mathbb{R}^2$ is the following one, which arose from work of Furstenberg.

For a given $\alpha \in (0, 1]$, suppose that E is a compact set in the plane, and for each $e \in S^1$ there is a line ℓ_e with direction e such that $\dim(\ell_e \cap E) \geq \alpha$. Then what is the smallest possible value for $\dim E$?

Easy results here are that $\dim E \geq \max(2\alpha, \frac{1}{2} + \alpha)$ and that there is an example with $\dim E = \frac{1}{2} + \frac{3}{2}\alpha$. We give proofs below. Several people have unpublished results on this question and it is unlikely that the author was the first to observe these bounds; in all probability they are due to Furstenberg and Katznelson.

The analogous discrete question is solved by the following result due to Szemerédi and Trotter [66] (see also [19], [50], [65]).

Suppose we are given n points $\{p_i\}$ and k lines $\{\ell_j\}$ in the plane. Define a line and point to be *incident*, $p \sim \ell$, if p lies on ℓ. Let $\mathcal{I} = \{(i, j) : p_i \sim \ell_j\}$. Then $|\mathcal{I}| \lesssim (kn)^{\frac{2}{3}} + k + n$, and this bound is sharp.

We note that the weaker bound $|\mathcal{I}| \lesssim (kn)^{\frac{3}{4}} + k + n$ follows from (166) and was known long before [66]. To see the analogy with the Hausdorff dimension question, reformulate the Szemerédi–Trotter bound as follows: if each line is incident to at least μ points ($\mu \gg 1$), then (since $|\mathcal{I}| \geq k\mu$)

$$(171) \qquad\qquad n \gtrsim \min(\mu^{\frac{3}{2}} k^{\frac{1}{2}}, \mu k).$$

Now assume say[4] that E has a covering by n discs D_i of radius δ. Consider a set of $k \approx \delta^{-1}$ δ-separated directions $\{e_j\}$. For each j the line ℓ_{e_j} will intersect D_i for at least $\delta^{-\alpha}$ values of i. We now pretend that we can replace points by the discs D_i in Szemerédi–Trotter and apply (171) with $\mu = \delta^{-\alpha}$, $k = \delta^{-1}$. Since $k \geq \mu$ we would obtain $n \gtrsim \delta^{-\frac{1}{2} - \frac{3}{2}\alpha}$, i.e. that the bound $\dim E \geq \frac{1}{2} + \frac{3}{2}\alpha$ should hold.

In one situation to be discussed in Section 3, it turns out that this kind of heuristic argument can be justified leading to a theorem in the continuum. In other situations such as the present one, it seems entirely unclear whether this should be the case or not, but still the discrete results suggest plausible conjectures.

If correct the bound $\dim E \geq \frac{1}{2} + \frac{3}{2}\alpha$ would be best possible by essentially the same example (due to Erdős, see [50]) that shows the Szemerédi–Trotter bound is sharp.

We start by recalling that if $\{n_j\}$ is a sequence of integers which increases sufficiently rapidly, and if $\alpha \in (0, 1)$ then the set

$$T \overset{\text{def}}{=} \left\{ x \in (\tfrac{1}{4}, \tfrac{3}{4}) : \forall j \; \exists p, q \in \mathbb{Z} : q \leq n_j^\alpha \text{ and } |x - \tfrac{p}{q}| \leq n_j^{-2} \right\}$$

has Hausdorff dimension α. This is a version of Jarnik's theorem—see [23, p. 134, Theorem 8.16(b)].

---

[4]In this heuristic argument we ignore the distinction between Hausdorff and Minkowski dimension.

It follows that the set

$$T' = \left\{ t : \frac{1-t}{t\sqrt{2}} \in T \right\}$$

also has dimension α.

For fixed n, consider the set of all line segments ℓ_{jk} connecting a point $(0, \frac{j}{n})$ to a point $(1, \frac{k}{n}\sqrt{2})$, where j and k are any integers between 0 and $n-1$. Thus $\ell_{jk} = \{(x, \phi_{jk}(x)) : 0 \le x \le 1\}$ where $\phi_{jk}(x) = (1-x)\frac{j}{n} + x\frac{k}{n}\sqrt{2}$. It follows using e.g. [**37**, p. 124, example 3.2] that every number in $[0, 1]$ differs by $\lesssim n^{-2}(\log n)^2$ from the slope of one of the ℓ_{jk}'s, so the set

$$G_n \overset{\text{def}}{=} \bigcup_{jk} S_{\ell_{jk}}^{n^{-2}(\log n)^3}$$

is a G-set.

Define

$$Q_n = \left\{ t : \frac{1-t}{t\sqrt{2}} \text{ is a rational number } \frac{p}{q} \in \left(\frac{1}{4}, \frac{3}{4}\right) \text{ with denominator } q \le n^\alpha \right\}.$$

If $t \in Q_n$, then let $S(t) \overset{\text{def}}{=} \{\phi_{jk}(t)\}_{j,k=0}^{n-1}$. For any j and k we have

$$(t\sqrt{2})^{-1}\phi_{jk}(t) = \frac{pj + qk}{qn},$$

a rational with denominator qn. We conclude that $|S(t)| \lesssim qn \le n^{1+\alpha}$, hence $\left|\bigcup(S(t) : t \in Q_n)\right| \lesssim n^{1+3\alpha}$ and

(*)   The set $\{(x, y) \in G_n : |x - t| \le \frac{1}{n^2} \text{ for some } t \in Q_n\}$ is contained in the union of $\lesssim n^{1+3\alpha}$ discs of radius $n^{-2}(\log n)^3$.

Now we let $\{n_j\}$ increase rapidly and will recursively construct compact sets F_j such that $F_{j+1} \subset F_j$, each F_j is a G-set and the set $\{(x, y) \in F_j : x \in T'\}$ is contained in the union of $n_j^{1+3\alpha} \log n_j$ discs of radius $n_j^{-2}(\log n_j)^3$. Namely, let F_0 be any G-set. If F_j has been constructed it will be of the form

$$\bigcup_{i=1}^{M} S_{\ell_i}^{\delta}$$

for a certain δ, where $\ell_i = \{(x, m_i x + b_i) : 0 \le x \le 1\}$ for suitable m_i and b_i, and every number in $[0, 1]$ is within δ of one of the m_i. As in the proof of Lemma 11.4 we let $A_i(x, y) = (x, m_i x + \delta y + b_i)$. We make n_{j+1} sufficiently large and define

$$F_{j+1} = \bigcup_{i=1}^{M} A_i(G_{n_{j+1}}).$$

Clearly $F_{j+1} \subset F_j$, and it follows as in Lemma 11.4 that the resulting set is a G-set. The covering property is also essentially obvious from (*) provided n_{j+1} is large enough, say $\log(n_{j+1}) \gg M$.

Let $F = \bigcap_j F_j$, and let $E = \{(x,y) \in F : x \in T'\}$. Then the covering property in the construction of F_j implies that $\dim E \leq \frac{1}{2}(1 + 3\alpha)$. On the other hand F is a G-set, and if ℓ is a line segment contained in F, then $\dim(\ell \cap E) = \dim T' \geq \alpha$. This completes the construction.

We now discuss the bound $\dim E \geq \max(2\alpha, \frac{1}{2}+\alpha)$. The bound $\dim E \geq 2\alpha$ can be derived from Proposition 11.8 by an argument like the proof of Lemma 11.9; we will omit this. To prove the bound $\dim E \geq \frac{1}{2} + \alpha$ (which corresponds to the easy $|\mathcal{I}| \lesssim (kn)^{\frac{3}{4}} + k + n$ under the above heuristic argument) fix a compact set E and for each $e \in S^1$ a line ℓ_e which intersects E in dimension $\geq \alpha$. Let $\{D_j\} = \{D(x_j, r_j)\}$ be a covering. Fix $\beta_1 < \beta < \alpha$; we have to bound $\sum_j r_j^{\frac{2}{2}+\beta_1}$ from below. As in the proof of Lemma 11.9 we let $\Sigma_k = \{j : 2^{-k} \leq r_j \leq 2^{-(k-1)}\}$, $\nu_k = |\Sigma_k|$ and $E_k = E \cap (\bigcup\{D_j : j \in \Sigma_k\})$. We start by choosing a number k and a subset $\Omega \subset S^1$ with measure $\gtrsim \frac{1}{k^2}$ such that if $e \in \Omega$ then $H_\beta(\ell_e \cap E_k) \geq C^{-1}k^{-2}$, using the pigeonhole principle as in the proof of Lemma 11.9. Let $\gamma = \frac{2}{\beta}$. Since $H_\beta(I) \leq |I|^\beta$ for any interval I, it follows that for a suitable numerical constant C, and for any $e \in \Omega$ there are *two* intervals $I_e^\pm$ on ℓ_e which are $C^{-1}k^{-\gamma}$- separated and such that $H_\beta(E_k \cap I_e^\pm) \gtrsim k^{-2}$. Let $\{e_i\}_{i=1}^M$ be a 2^{-k}-separated subset of Ω with $M \gtrsim \frac{2^k}{k^2}$ (see (169)) and define
(172)
$$\mathcal{T} = \{(j_+, j_-, i) \in \Sigma_k \times \Sigma_k \times \{1, \ldots, M\} : I_{e_i}^+ \cap E_k \cap D_{j_+} \neq \emptyset, I_{e_i}^- \cap E_k \cap D_{j_-} \neq \emptyset\}.$$

We will count $\mathcal{T}$ in two different ways.

First fix j_+ and j_- and consider how many values of i there can be with $(j_+, j_-, i) \in \mathcal{T}$. We will call such a value of i *allowable*. If the distance between D_{j_+} and D_{j_-} is small compared with $k^{-\gamma}$ then there is no allowable i, since the distance between $I_{e_i}^+$ and $I_{e_i}^-$ is always $\geq C^{-1}k^{-\gamma}$. On the other hand if the distance between D_{j_+} and D_{j_-} is $\gtrsim k^{-\gamma}$, then because the $\{e_i\}$ are 2^{-k}-separated, it follows that there are $\lesssim k^\gamma$ i's such that ℓ_{e_i} intersects both D_{j_+} and D_{j_-}. Hence in either case there are $\lesssim k^\gamma$ allowable i's. Summing over j_+ and j_- we conclude that

(173)
$$|\mathcal{T}| \lesssim k^\gamma \nu_k^2.$$

On the other hand, for any fixed i, the lower bound $H_\beta(E_k \cap I_{e_i}^+) \gtrsim k^{-2}$ implies there are $\gtrsim k^{-\gamma}2^{k\beta}$ values of j_+ such that $I_{e_i}^+ \cap E_k \cap D_{j_+} \neq \emptyset$ and similarly with $+$ replaced by $-$. So $|\mathcal{T}| \gtrsim M(k^{-\gamma}2^{k\beta})^2$. Comparing this bound with (173) we conclude that

$$\nu_k \gtrsim k^{-\frac{3}{2}\gamma}2^{k\beta}\sqrt{M} \gtrsim k^{-(1+\frac{3}{2}\gamma)}2^{(\frac{1}{2}+\beta)k} \gtrsim 2^{(\frac{1}{2}+\beta_1)k}$$

and therefore $\sum_{j \in \Sigma_k} r_j^{\frac{1}{2}+\beta_1} \geq \text{constant}$. $\qquad\square$

11.2. The higher dimensional case

We will first make a few remarks about the corresponding problem over finite fields, which is the following:

Let $\mathbb{F}_q$ be the field with q elements and let V be an n-dimensional vector space over $\mathbb{F}_q$. Let E be a subset of V which contains a line in every direction, i.e.

$$\forall e \in V \setminus \{0\} \ \exists a \in V : a + te \in E \ \forall t \in \mathbb{F}_q.$$

Does it follow that $|E| \geq C_n^{-1} q^n$?

Of course C_n should be independent of q. One could ask instead for a bound like $\forall \epsilon > 0 \ \exists C_{n\epsilon} : |E| \geq C_{n\epsilon}^{-1} q^{n-\epsilon}$ or could restrict to the case of prime fields $\mathbb{F}_p$ or fields with bounded degree over the prime field.

So far as I have been able to find out this question has not been considered, and the simple result below corresponds to what is known in the Euclidean case[5].

PROPOSITION 11.12. *In the above situation* $|E| \geq C_n^{-1} q^{\frac{n+2}{2}}$.

We give the proof since it is based on the same idea as the $\mathbb{R}^n$ proof but involves no technicalities.

First consider the case $n = 2$, which is analogous to Proposition 11.8. We will actually prove the following more general statement, which we need below: suppose (with $\dim V = 2$) that E contains at least $\frac{q}{2}$ points on a line in each of m different directions. Then

$$(174) \qquad\qquad |E| \gtrsim mq.$$

To prove (174), let $\{\ell_j\}_{j=1}^m$ be the lines. Any two distinct ℓ_j's intersect in a point. Accordingly

$$\tfrac{1}{2}qm \leq \sum_j |E \cap \ell_j| \leq |E|^{\frac{1}{2}} \left(\sum_{jk} |\ell_j \cap \ell_k| \right)^{\frac{1}{2}} = |E|^{\frac{1}{2}} (m(m-1+q))^{\frac{1}{2}} \leq |E|^{\frac{1}{2}} (mq)^{\frac{1}{2}}$$

where we used that $m \leq q+1$. It follows that $|E| \gtrsim mq$. Taking $m = q+1$ we obtain the two dimensional case of Proposition 11.12.

Now assume $n \geq 3$. Then E contains $\frac{q^n-1}{q-1} \approx q^{n-1}$ lines $\{\ell_j\}$. Fix a number μ and define a *high multiplicity line* to be a line ℓ_k with the following property: for at least $\frac{q}{2}$ of the q points $x \in \ell_k$, the set $\{j : x \in \ell_j\}$ has cardinality at least $\mu + 1$. Consider two cases: (i) no high multiplicity line exists (ii) a high multiplicity line exists.

In case (i) we define $\tilde{E} = \{x \in E : x \text{ belongs to } \leq \mu \ \ell_j\text{'s}\}$. Then $\tilde{E}$ intersects each ℓ_j in at least $\frac{q}{2}$ points, by definition of case (i). Each point of $\tilde{E}$ belongs to at most $\mu \ \ell_j$'s so we may conclude that

$$|E| \geq |\tilde{E}| \geq \mu^{-1} \sum_j |\tilde{E} \cap \ell_j| \gtrsim \mu^{-1} q \cdot q^{n-1}.$$

In case (ii), let $\{\Pi_i\}$ be an enumeration of the 2-planes containing ℓ_k. By definition of high multiplicity line there are at least $\frac{\mu q}{2}$ lines ℓ_j, $j \neq k$, which

---

[5]*Editor's note:* The finite field Kakeya problem posed here was subsequently studied further in [46], [71], [14]. There has also been significant progress on the Euclidean case, see the footnote to Remark 2.2.

intersect ℓ_k. Each one of them is contained in a unique Π_i, and contains $q-1$ points of Π_i which do not lie on ℓ_k. Let $\mathcal{L}_i$ be the set of lines ℓ_j which are contained in a given Π_i. Then by (174) we have $\left|E \cap \Pi_i \cap (V\backslash\ell_k)\right| \gtrsim q|\mathcal{L}_i|$. The sets $\Pi_i \cap (V\backslash\ell_k)$ are pairwise disjoint so we can sum over i to get $|E| \gtrsim q\sum_i |\mathcal{L}_i| \geq \frac{q^2\mu}{2}$.

If we take μ to be roughly $q^{\frac{n-2}{2}}$ we obtain $|E| \gtrsim q^{\frac{n+2}{2}}$ in either case (i) or (ii), hence the result. $\qquad\Box$

REMARK 11.13. General finite fields do not always resemble the Euclidean case in this sort of problem. For example, the Szemerédi–Trotter theorem is easily seen to be false (e.g. [**5**, p. 75]). A counterexample involving one line in each direction as in Remark 11.11 may be obtained in the following way: let $q = p^2$ with p prime, let α be a generator of $\mathbb{F}_q$ over $\mathbb{F}_p$ and in the two dimensional vector space V over $\mathbb{F}_q$, let ℓ_{jk} be the line connecting $(0, j)$ to $(1, k\alpha)$. Here j and k are in $\mathbb{F}_p$. This is a set of p^2 lines containing one line in each direction other than the vertical. For given $t \in \mathbb{F}_q$, let $S_t = \{y \in \mathbb{F}_q : (t, y) \in \bigcup_{jk} \ell_{jk}\}$. If t is such that $\alpha\frac{t}{1-t} \in \mathbb{F}_p$ then it is easily seen that S_t coincides with $(1-t)\mathcal{F}_p$, and if $t = 1$ then $S_t = \alpha\mathbb{F}_p$. This gives p "bad" values of t such that S_t has cardinality p. Let $E = \bigcup_t \{(t, y) : y \in S_t\}$, where the union is taken over the bad values of t. Then $\{\ell_{jk}\}$ and E give a configuration of p^2 lines and p^2 points with p^3 incidences, matching the trivial upper bound from (166).

In the $\mathbb{R}^n$ context, arguments like the proof of Proposition 11.12 can still be used, except that one has to work with tubes instead of lines and measure instead of cardinality, and take into account such issues as that the size of the intersection of two tubes will depend on the angle of intersection via (167). This was perhaps first done by Córdoba (e.g. [**20**]—see the proof of Proposition 11.8 above). We will present here the "bush" argument from [**7**, p. 153-4], which shows the following:

PROPOSITION 11.14. $\|f_\delta^*\|_{n+1,\infty} \leq C_n \delta^{-\frac{n-1}{n+1}} \|f\|_{\frac{n+1}{2},1}$.

PROOF. Using (169), we see that what must be shown is the following: if $\{T_{e_j}^\delta\}_{j=1}^M$ are tubes with δ-separated directions, E is a set and $|E \cap T_{e_j}| \geq \lambda |T_{e_j}^\delta|$, then

$$(175) \qquad\qquad |E| \gtrsim \delta^{\frac{n-1}{2}} \lambda^{\frac{n+1}{2}} \sqrt{M}.$$

To this end we fix a number μ ("multiplicity") and consider the following two possibilities:

(i) (low multiplicity) No point of E belongs to more than μ tubes $T_{e_j}^\delta$.

(ii) (high multiplicity) Some point $a \in E$ belongs to more than μ tubes $T_{e_j}^\delta$.

In case (i) it is clear that $|E| \gtrsim \mu^{-1} \sum_j |E \cap T_{e_j}^\delta|$, hence

$$(176) \qquad\qquad |E| \gtrsim \mu^{-1} M \lambda \delta^{n-1}.$$

In case (ii) we fix a point a as indicated and may assume that a belongs to $T_{e_j}^\delta$ when $j \le \mu + 1$. If C_0 is a suitably large fixed constant, then $\left| T_{e_j}^\delta \cap D(a, C_0^{-1}\lambda) \right| \le \frac{1}{2} |T_{e_j}^\delta|$. Accordingly, for $j \le \mu + 1$, we have

$$\left| E \cap T_{e_j}^\delta \cap D(a, C_0^{-1}\lambda)^c \right| \ge \frac{\lambda}{2} |T_{e_j}^\delta| \gtrsim \lambda \, \delta^{n-1}.$$

If $j, k \le \mu$ then $T_{e_j}^\delta \cap T_{e_k}^\delta$ contains a and has diameter $\lesssim \frac{\delta}{\theta(e_j, e_k)}$ by (167). It follows that if $\theta(e_j, e_k) \ge C_1 \frac{\delta}{\lambda}$ for a suitably large C_1, then the sets $E \cap T_{e_j}^\delta \cap D(a, C_0^{-1}\lambda)^c$ and $E \cap T_{e_k}^\delta \cap D(a, C_0^{-1}\lambda)^c$ are disjoint. We conclude that

$$|E| \gtrsim \mathcal{N} \cdot \lambda \, \delta^{n-1}$$

where $\mathcal{N}$ is the maximum possible cardinality for a $C_1 \frac{\delta}{\lambda}$-separated subset of $\{e_j\}_{j=1}^{\mu+1}$. Since the $\{e_j\}$ are δ- separated, we have $\mathcal{N} \gtrsim \lambda^{n-1}\mu$ and therefore

$$(177) \qquad\qquad |E| \gtrsim \lambda^n \delta^{n-1}\mu.$$

We conclude that for any given μ either (176) or (177) must hold. Taking $\mu \approx \lambda^{-(\frac{n-1}{2})}\sqrt{M}$ we get (175). $\qquad\square$

Further remarks.

REMARK 11.15. Bourgain [7] also gave an additional argument leading to an improved result which implies $\dim(\text{Kakeya}) \ge \frac{n+1}{2} + \epsilon_n$, where ϵ_n is given by a certain inductive formula (in particular $\epsilon_3 = \frac{1}{3}$). A more efficient argument was then given by the author [73], based on considering families of tubes which intersect a line instead of a point as in the bush argument; this is the continuum analogue of the proof of Proposition 11.12. It gives the bound

$$(178) \qquad\qquad \forall \epsilon \; \exists C_\epsilon : \|f_\delta^*\|_q \le C_\epsilon \delta^{-(\frac{n}{p}-1)-\epsilon}\|f\|_p,$$

where $p = \frac{n+2}{2}$ and $q = (n-1)p'$. This is the estimate on L^p which would follow by interpolation with the trivial $\|f_\delta^*\|_\infty \lesssim \delta^{-(n-1)}\|f\|_1$ if the bound (165) could be proved. In particular, it implies the dimension of Kakeya sets is $\ge \frac{n+2}{2}$. Other proofs of estimates like (178) have also recently been given by Katz (see [46]) and Schlag [55]. However in every dimension $n \ge 3$ it is unknown whether (178) holds for any $p > \frac{n+2}{2}$ and whether $\dim(\text{Kakeya}) > \frac{n+2}{2}$[6].

REMARK 11.16. Proposition 11.14 is also a corollary of the $L^{\frac{n+1}{2}} \to L^{n+1}$ estimate for the x-ray transform due to Drury and Christ [22], [20] (see also [49], [18] for related results). Conversely, a refinement of the argument which proves (178) can be used to prove the estimate on $L^{\frac{n+2}{2}}$ which would

---

[6]*Editor's note:* Such improvements have since been obtained by Bourgain, Katz, Laba and Tao [13], [31], [32], [33], [39].

follow from (165) and the result of [22] by interpolation, at least in the three dimensional case. See [75][7].

REMARK 11.17. We briefly discuss some other related problems. The classical problem of Nikodym sets has been shown to be formally equivalent to the Kakeya problem by Tao [68]; we refer to his paper for further discussion. Another classical problem is the problem of $(n, 2)$ sets: suppose that E is a set in $\mathbb{R}^n$ which contains a translate of every 2-plane. Does it follow that E has positive measure? At present this is known only when $n = 3$ [41] or $n = 4$ [7]. The argument in [7], Section 4 shows the following: suppose that (165) can be proved in dimension $n - 1$, or more precisely that a slightly weaker result can be proved, namely that for some p and q there is an estimate

$$(179) \qquad \|f_\delta^*\|_{L^q(S^{n-2})} \lesssim \delta^{-\alpha} \|f\|_{L^p(\mathbb{R}^{n-1})} \text{ with } \alpha < \tfrac{1}{p}.$$

Then $(n, 2)$-sets have positive measure.

However, note that (179) would imply by Lemma 11.9 that Kakeya sets in $\mathbb{R}^{n-1}$ have dimension $\geq n - 2$. In fact if an estimate (179) is true for every n then one could answer Question 1 affirmatively by an argument based on the fact that the direct product of Kakeya sets is Kakeya. It may therefore be unlikely that the $(n, 2)$-sets problem can be solved without a full understanding of the Kakeya problem. However, the most recent results on it are those of [1][8].

REMARK 11.18. If one considers curves instead of lines, then it is known that much less can be expected to be true. This first results in this direction are in [8]; see also [10], [43] and [59].

11.3. Circles

In this section we will discuss some analogous problems about circles in the plane, or (essentially equivalent) fine estimates for the wave equation in $2 + 1$ dimensions. These problems are much better understood than the Kakeya problem and yet they present some of the same difficulties.

A prototype result due to Bourgain [6] and Marstrand [42] independently is that

($*$) A set in $\mathbb{R}^2$ which contains circles with arbitrary centers must have positive measure.

Bourgain proved a stronger result which has the same relation to ($*$) as Question 2 does to Question 1. Namely, define a maximal function

$$\mathcal{M}f(x) = \sup_r \int \left| f(x + re^{i\theta}) \right| \frac{d\theta}{2\pi}.$$

---

[7]*Editor's note:* See also [38] for a higher-dimensional analogue.

[8]*Editor's note:* More recently, Mitsis [44] proved that $(n, 2)$-sers in $\mathbb{R}^n$, $n > 4$, have Hausdorff dimension n.

Then $\|\mathcal{M}f\|_{L^p(\mathbb{R}^2)} \lesssim \|f\|_{L^p(\mathbb{R}^2)}$, $p > 2$. As is well-known, this maximal function was introduced by Stein [61] and he proved the analogous inequality in dimensions $n \geq 3$; the range of p is then $p > \frac{n}{n-1}$. Stein's proof was based partly on the Plancherel theorem and Bourgain's argument in the two dimensional case also used the Plancherel theorem, whereas Marstrand's argument was purely geometric. We will discuss some further developments of the latter approach.

A variant on the Kakeya construction due to Besicovitch–Radó [4] and Kinney [35] shows the following:

(∗∗) There are compact sets in the plane with measure zero containing circles of every radius between 1 and 2.

We will call such sets BRK sets. The distinction between (∗) and (∗∗) can be understood in terms of parameter counting: a set as in (∗∗) is a subset of a 2-dimensional space containing a 1-parameter family of 1-dimensional objects, so whether it has positive measure or not can be expected to be a borderline question. This is analogous to the question of Kakeya sets which also contain $n - 1$-parameter families of 1-dimensional objects. On the other hand a set as in (∗) contains a 2- parameter family of 1-dimensional objects in a 2-dimensional space.

A further related remark is that analogous constructions with other 1-parameter families of circles have been done by Talagrand [67]. For example, he shows that for any smooth curve γ there are sets of measure zero containing circles centered at all points of γ.

It is natural to ask whether the dimension of a BRK set must be 2 or not. This question also has a maximal function version; the relevant maximal function is the following M_δ: if $f : \mathbb{R}^2 \to \mathbb{R}$ then $M_\delta f : [\frac{1}{2}, 2] \to \mathbb{R}$,

$$(180) \qquad M_\delta f(r) = \sup_x \frac{1}{|C_\delta(x,r)|} \int_{C_\delta(x,r)} |f|.$$

One shows analogously to Lemma 11.9 that a bound (for some $p < \infty$)

$$(181) \qquad \forall \epsilon \; \exists \mathbb{C}_\epsilon : \|M_\delta f\|_{L^p([\frac{1}{2},2])} \leq C_\epsilon \delta^{-\epsilon} \|f\|_p$$

will imply that BRK sets have dimension 2. Note that existence of measure zero BRK sets implies the $\delta^{-\epsilon}$ factor is needed. This is similar to the situation with the two dimensional Kakeya problem. However in contrast to the latter problem it is not possible to take $p = 2$ in (181). In fact p must be at least 3; this is seen by considering the standard example $f =$ indicator function of a rectangle with dimensions $\delta \times \sqrt{\delta}$.

REMARK 11.19. Sets in $\mathbb{R}^n$ with measure zero containing spheres of all radii may be shown to exist for $n \geq 3$ also, and the maximal function (180) may be defined in $\mathbb{R}^n$. However, in that case the questions mentioned above are essentially trivial, since the correct estimate for the maximal function is an $L^2 \to L^2$ estimate, is easy and implies that sets containing spheres with all radii have dimension n. Namely, the estimate

$$(182) \qquad \|M_\delta f\|_2 \lesssim \left(\log\tfrac{1}{\delta}\right)^{\frac{1}{2}} \|f\|_2$$

can be proved analogously to Proposition 11.8 and is also closely related to some of the Strichartz inequalities for the wave equation (cf. [51]), due to the fact that spherical means correspond roughly to solutions of the initial value problem $\Box u = 0$, $u(\,\cdot\,,0) = f$, $\frac{\partial u}{\partial t}(\,\cdot\,,0) = 0$ after taking $\frac{n-1}{2}$ derivatives. These remarks are from [36]. From a certain point of view, the "reason" why the higher dimensional case is easier is the following: if $|r - s| \approx 1$ then

(183)

$$|C_\delta(x,r) \cap C_\delta(y,s)| \approx \begin{cases} \delta^{\frac{n+1}{2}} & \text{if } C(x,r) \text{ and } C(y,s) \text{ are tangent} \\[2mm] \delta^2 & \text{if } C(x,r) \text{ and } C(y,s) \text{ are sufficiently transverse} \end{cases}$$

making the first possibility "worse" than the second in $\mathbb{R}^2$ but not in higher dimensions.

We now consider only the two dimensional case and will formulate a discrete analogy like the analogy between the Szemerédi–Trotter theorem and the question mentioned in Remark 11.11 The relevant problem in discrete geometry is

Given N circles $\{C_i\}$ in the plane, no three tangent at a point, how many pairs (i,j) can there be such that C_i is tangent to C_j?

For technical reasons we always interpret "tangent" as meaning "internally tangent", i.e. a circle $C(x,r)$ is "tangent" to $C(y,s)$, written $C(x,r)\|C(y,s)$, iff $|x - y| = |r - s|$.

We will call this the tangency counting problem. We're not aware of any literature specifically about this problem, but known techniques in incidence geometry (related to the Szemerédi–Trotter theorem) can be adapted to it without difficulty. One obtains the following bounds for $\mathcal{I} \overset{\text{def}}{=} \{(i,j) : C_i\|C_j\}$.

(i) (easy) $|\mathcal{I}| \lesssim N^{\frac{5}{3}}$. This follows from the fact that the incidence matrix

$$a_{ij} = \begin{cases} 1 & \text{if } C_i\|C_j \\ 0 & \text{otherwise} \end{cases}$$

contains no 3×3 submatrix of 1's (essentially a theorem of Apollonius: there are at most two circles which are internally tangent to three given circles at distinct points) and therefore contains at most $\mathcal{O}(N^{\frac{5}{3}})$ 1's by (166).

(ii) (more sophisticated) $\forall \epsilon > 0\ \exists C_\epsilon < \infty : |\mathcal{I}| \lesssim N^{\frac{3}{2}+\epsilon}$. This follows readily from the techniques of Clarkson, Edelsbrunner, Guibas, Sharir and Welzl [19]. We will not discuss their work here; we just note that they prove the analogous $N^{\frac{3}{2}+\epsilon}$ bound in the three dimensional unit distance problem: in our notation, given $\{(x_i,r_i)\}_{i=1}^{N} \subset \mathbb{R}^2 \times \mathbb{R}$, there are $\lesssim N^{\frac{3}{2}+\epsilon}$ pairs (i,j) with $|x_i - x_j|^2 + (r_i - r_j)^2 = 1$.

There is no reason to think that the bound (ii) should be sharp.[9] However, (ii) leads to a sharp result on the BRK sets problem and a proof of the maximal inequality (181) with $p = 3$. The heuristic argument is the following: assume we know a bound $\lesssim N^\alpha$ in the tangency counting problem, where $\alpha \geq \frac{3}{2}$. Let E be a BRK set and consider its δ-neighborhood E^δ. Let $\{r_j\}_{j=1}^{M}$ be a maximal δ-separated subset of $[\frac{1}{2}, 2]$; then $M \approx \frac{1}{\delta}$ and E^δ contains an annulus $C_\delta(x_j, r_j)$ for each j. By (183), we should have to a first approximation $|C_\delta(x_j, r_j) \cap C_\delta(x_k, r_k)| \approx \delta^{\frac{3}{2}}$ if $C(x_j, r_j)$ and $C(x_i, r_i)$ intersect tangentially and $|C_\delta(x_j, r_j) \cap C_\delta(x_k, r_k)| \approx \delta^2$ if they intersect transversally. Accordingly we would get

$$\sum_{jk} \left| C_\delta(x_j, r_j) \cap C_\delta(x_k, r_k) \right| \lesssim \delta^{-\alpha} \cdot \delta^{\frac{3}{2}} + \delta^{-2} \cdot \delta^2 \lesssim \delta^{\frac{3}{2} - \alpha},$$

and then the argument in the proof of Proposition 11.8 shows that $|E_\delta| \gtrsim \delta^{\frac{1}{2}(\alpha - \frac{3}{2})}$, so one expects $\dim E \geq 2 - \frac{1}{2}(\alpha - \frac{3}{2})$.

It turns out that it is possible to make this argument rigorous and to obtain a corresponding result ((181) with $p = 3$) for the maximal operator. The first lemma below keeps track of the intersection of two annuli in terms of their degree of tangency; it is of course quite standard and is used in one form or another in most papers in the area, e.g. [6] and [42]. The second lemma is due to Marstrand ([42], Lemma 5.2), although he formulated it slightly differently. It gives a quantitative meaning to the theorem of Apollonius used in the proof of the $N^{\frac{5}{3}}$ tangency bound.

We introduce the following notation: if $C(x, r)$ and $C(y, s)$ are circles then

$$d((x, r), (y, s)) = |x - y| + |r - s|,$$
$$\Delta((x, r), (y, s)) = \big| |x - y| - |r - s| \big|.$$

Note that Δ vanishes precisely when the circles are "tangent." In Lemmas 11.20 and 11.21 below, we assume that all circles $C(x, r)$ etc. have centers in $D(0, \frac{1}{4})$ and radii between $\frac{1}{2}$ and 2.

LEMMA 11.20. *Assume that* $x \neq y$. *Let* $d = d((x, r), (y, s))$, $\Delta = \Delta((x, r), (y, s))$, *and* $e = sgn(r - s)\frac{y-x}{|y-x|}$, $\zeta = y + re$. *Then*

(a) $C_\delta(x, r) \cap C_\delta(y, s)$ *is of measure* $\lesssim \delta \cdot \dfrac{\delta}{\sqrt{(\delta + \Delta)(\delta + d)}}$.

(b) $C_\delta(x, r) \cap C_\delta(y, s)$ *is contained in a disc centered at* ζ *with radius* $\lesssim \sqrt{\dfrac{\Delta + \delta}{d + \delta}}$.

---

[9]It may be more natural to consider a slightly different formulation of the problem: drop the assumption that no three circles are tangent at a point, and consider the number of points where two are tangent instead of the number of tangencies. With this reformulation, a standard example involving circles with integer center and radius shows that the exponent $\frac{4}{3}$ would be best possible as in the unit distance problem.

PROOF. We use the following fact: if $\mu > 0, \epsilon > 0$ then the set

$$\{x \in [-\pi, \pi] : |\cos x - \mu| \leq \epsilon\}$$

is (i) contained in the union of two intervals of length $\lesssim \dfrac{\epsilon}{\sqrt{|1-\mu|}}$ and (ii) contained in an interval of length $\lesssim \sqrt{|1 - \mu| + \epsilon}$ centered at 0.

To prove the lemma, we use complex notation and may assume that $x = 0$, $r = 1$, y is on the positive real axis and $s < 1$. Note that then $e = 1$. If $d \leq 4\delta$ then the lemma is trivial, and if $y < \frac{d}{2} - \delta$ then $y + s < 1 - 2\delta$ so that $C_\delta(0, 1) \cap C_\delta(y, s) = \emptyset$. So we can assume that $d \geq 4\delta$ and $y \geq \frac{d}{2} - \delta \geq \frac{d}{4}$.

If $z \in C_\delta(0, 1) \cap C_\delta(y, s)$ then clearly $|z - e^{i\theta}| \leq \delta$ for some $\theta \in [-\pi, \pi]$. It suffices to show that the set of θ which can occur here is contained in two intervals of length $\lesssim \dfrac{\delta}{\sqrt{(\delta + \Delta)(\delta + d)}}$ and in an interval of length $\lesssim \sqrt{\dfrac{\Delta + \delta}{d + \delta}}$ centered at 0.

The point $e^{i\theta}$ must belong to $C_{2\delta}(y, s)$, i.e. $\left| |e^{i\theta} - y| - s \right| < 2\delta$ and therefore, since $\left| |e^{i\theta} - y| + s \right| \approx 1$,

$$\left| |e^{i\theta} - y|^2 - s^2 \right| \lesssim \delta.$$

We can express this as

$$\left| \cos\theta - \frac{1 + y^2 - s^2}{2y} \right| \lesssim \frac{\delta}{y} \sim \frac{\delta}{d}.$$

Let $\mu = \dfrac{1 + y^2 - s^2}{2y}$, $\epsilon = C\dfrac{\delta}{d}$. Then μ is positive, and

$$|1 - \mu| = \frac{|s^2 - (1 - y)^2|}{2y} \approx \frac{|1 - s - y|}{2y} \approx \frac{\Delta}{d}.$$

Apply fact (ii) in the first paragraph. The set of possible θ is therefore contained in an interval of length $\lesssim \sqrt{\dfrac{\Delta + \delta}{d}}$ centered at 0. This proves (b), since we are assuming $d \geq \delta$. Estimate (a) follows from (b) if $\Delta \leq \delta$. If $\Delta \geq \delta$, then fact (i) in the first paragraph gives the additional property that θ must be contained in the union of two intervals of length $\lesssim \dfrac{\delta/d}{\sqrt{\Delta/d}} \approx \dfrac{\delta}{\sqrt{(\delta + \Delta)(\delta + d)}}$. $\qquad\square$

LEMMA 11.21 (Marstrand's 3-circle lemma). *For a suitable numerical constant C_0, assume that $\epsilon, t, \lambda \in (0, 1)$ satisfy $C_0 \frac{\epsilon}{t} \leq \lambda^2$. Fix three circles $C(x_i, r_i), 1 \leq i \leq 3$. Then for $\delta \leq \epsilon$ the set*

$$\overline{\Omega}_{\epsilon t \lambda} \overset{\text{def}}{=} \{(x, r) \in \mathbb{R}^2 \times \mathbb{R} : \Delta((x, r), (x_i, r_i)) < \epsilon \; \forall i,$$
$$d((x, r), (x_i, r_i)) > t \; \forall i, \; C_\delta(x, r) \cap C_\delta(x_i, r_i) \neq \emptyset \; \forall i,$$
$$\text{dist}(C_\delta(x, r) \cap C_\delta(x_i, r_i), C_\delta(x, r) \cap C_\delta(x_j, r_j)) \geq \lambda \; \forall i, j : i \neq j\}$$

is contained in the union of two ellipsoids in $\mathbb{R}^3$ each of diameter $\lesssim \frac{\epsilon}{\lambda^2}$ and volume $\lesssim \frac{\epsilon^3}{\lambda^3}$.

PROOF. This will be based on the inverse function theorem. We remark that the sketch of proof given in [**74**] is inaccurate.

We will actually work with a slightly different set, namely, with

$$\Omega_{\epsilon t \lambda} \;=\; \{(x,r) \in \mathbb{R}^2 \times \mathbb{R} : \Delta((x,r),(x_i,r_i)) < \epsilon \; \forall i,$$
$$d((x,r),(x_i,r_i)) > t \; \forall i, |e_i(x,r) - e_j(x,r)| \geq \lambda \; \forall i,j : i \neq j \},$$

where $e_i(x,r) = \operatorname{sgn}(r-r_i)\frac{x_i - x}{|x_i - x|}$. This is sufficient since by Lemma 11.20 (b), $\Omega_{\epsilon t \frac{\lambda}{2}}$ will contain $\overline{\Omega}_{\epsilon t \lambda}$ provided C_0 is sufficiently large.

If $e_1, \ldots, e_4$ are unit vectors in $\mathbb{R}^2$ which are contained in an arc of length μ, then the reader will convince herself or himself that

$$(184) \qquad\qquad |(e_1 - e_2) \wedge (e_3 - e_4)| \lesssim \mu |e_1 - e_2| \, |e_3 - e_4|$$

and furthermore if e_1, e_2, e_3 are unit vectors in $\mathbb{R}^2$ then

$$(185) \qquad |(e_1 - e_2) \wedge (e_1 - e_3)| \approx |e_1 - e_2| \, |e_2 - e_3| \, |e_3 - e_1|.$$

Here $\wedge$ is wedge product, $(a,b) \wedge (c,d) = ad - bc$.

Consider the map $G : \mathbb{R}^2 \times \mathbb{R} \to \mathbb{R}^3$ defined by

$$G(x,\rho) = \begin{pmatrix} |x - x_1| - |r - r_1| \\ |x - x_2| - |r - r_2| \\ |x - x_3| - |r - r_3| \end{pmatrix}.$$

Fix $(\xi,\rho) \in \Omega_{\epsilon t \lambda}$. Observe that

$$(186) \qquad\qquad DG(\xi,\rho) \approx \begin{pmatrix} e_1(\xi,\rho) & -1 \\ e_2(\xi,\rho) & -1 \\ e_3(\xi,\rho) & -1 \end{pmatrix},$$

where "$\approx$" means that the two matrices are equal after each row of the matrix on the right hand side is multiplied by an appropriate choice of ± 1.

We can assume that

$$|e_1(\xi,\rho) - e_3(\xi,\rho)| \geq |e_1(\xi,\rho) - e_2(\xi,\rho)| \geq |e_2(\xi,\rho) - e_3(\xi,\rho)|.$$

Let $\mu = |e_1(\xi,\rho) - e_3(\xi,\rho)|$, $\nu = |e_2(\xi,\rho) - e_3(\xi,\rho)|$; then we have $\mu \geq \nu \gtrsim \lambda$ and also $|e_1(\xi,\rho) - e_2(\xi,\rho)| \approx \mu$. It follows by (185) that $|\det DG(\xi,\rho)| \approx \mu^2 \nu$. Furthermore, all entries in the cofactor matrix of $DG(\xi,\rho)$ are easily seen to be $\lesssim \mu$. Let $E(\xi,\rho) = \{(x,r) \in \mathbb{R}^2 \times \mathbb{R} : |DG(\xi,\rho)(x-\xi, r-\rho)| < A\epsilon\}$ for an appropriate large constant A which should be chosen before C_0. Then the preceding considerations imply $E(\xi,\rho)$ is an ellipsoid with

$$(187) \qquad\qquad \operatorname{diam}(E(\xi,\rho)) \;\lesssim\; \frac{\epsilon}{\mu\nu},$$

$$(188) \qquad\qquad |E(\xi,\rho)| \;\lesssim\; \frac{\epsilon^3}{\mu^2 \nu}.$$

We claim that if $(x,r) \in E$ then $DG(x,r)DG(\xi,\rho)^{-1} = I + E$, where I is the 3×3 identity matrix and E is a matrix with norm $\leq \frac{1}{100}$, say.

A matrix calculation shows that each entry of $(DG(x,r) - DG(\xi,\rho))DG(\xi,\rho)^{-1}$ has the form $(\det DG(\xi,\rho))^{-1}(e_i(x,r) - e_i(\xi,\rho)) \wedge (e_j(\xi,\rho) - e_k(\xi,\rho))$ for appropriate i,j,k. We will show below that

$$(189) \qquad\qquad |e_i(x,r) - e_i(\xi,\rho)| \lesssim \frac{\epsilon}{t\nu}.$$

If we assume this then the claim may be proved as follows. (189) implies in particular that all the vectors $e_i(x,r)$ and $e_j(\xi,\rho)$ belong to an arc of length $\lesssim \mu$. Accordingly, using (184),

$$|\det DG(\xi,\rho)^{-1}(e_i(x,r) - e_i(\xi,\rho)) \wedge (e_j(\xi,\rho) - e_k(\xi,\rho))|$$
$$\lesssim \mu \left|\det DG(\xi,\rho)^{-1}\right| \left|e_i(x,r) - e_i(\xi,\rho)\right| \left|e_j(\xi,\rho) - e_k(\xi,\rho)\right|$$
$$\lesssim \mu \cdot (\mu^2\nu)^{-1} \cdot \frac{\epsilon}{t\nu} \cdot \mu \le \frac{\epsilon}{t\nu^2},$$

which is small.

To prove (189) we abbreviate $e_i = e_i(\xi,\rho)$. Fix i and let $e_i^* \in \mathbb{R}^2$ be a unit vector perpendicular to e_i. If we define j and k via $\{i,j,k\} = \{1,2,3\}$, then a little linear algebra shows that $e_i^* = \alpha(e_i - e_j) + \beta(e_i - e_k)$ with $|\alpha| + |\beta| \lesssim \nu^{-1}$. Furthermore, if we let $(v_1, v_2, v_3) = DG(\xi,\rho)(x - \xi, r - \rho)$, then by (186) we have $|(e_i - e_j)\cdot(x - \xi)| = |v_i \pm v_j| \le 2\epsilon$ and similarly $|(e_i - e_k)\cdot(x - \xi)| \le 2\epsilon$. We conclude that $|e_i^* \cdot (x - \xi)| \lesssim \frac{\epsilon}{\nu}$, hence $|e_i^* \cdot (x - x_i)| \lesssim \frac{\epsilon}{\nu}$ since $x_i - \xi$ is parallel to e_i. Also $|x - x_i| \ge \frac{t}{2}$ by (187), so

$$\left| e_i^* \cdot \frac{x - x_i}{|x - x_i|} \right| \lesssim \frac{\epsilon}{t\nu}.$$

This implies that for an appropriate choice of $\pm$

$$(190) \qquad\qquad |e_i(x,r) \pm e_i| \lesssim \frac{\epsilon}{t\nu}.$$

Note though that $r - r_i$ and $|x - x_i|$ are nonzero on $E(\xi,\rho)$: this follows from (187), since ϵ is small compared with t so that $|\xi - x_i| \approx t \approx |\rho - r_i|$. So $(x,r) \to e_i(x,r)$ is a continuous function on $E(\xi,\rho)$ and therefore the sign in (190) is independent of (x,r). So (189) holds and the claim is proved.

If A is large enough then the claim implies via the usual proof of the inverse function theorem that G is a diffeomorphism from a subset of $E(\xi,\rho)$ onto a disc of radius 2ϵ, say. In particular, $E(\xi,\rho)$ must contain a point (x,r) with $G(x,r) = 0$. Then $C(x,r)$ is internally tangent to each $C(x_i,r_i)$; note that by (189) and the bound on the diameter of E, we have $(x,r) \in \Omega_{\epsilon\frac{t}{2}\frac{\lambda}{2}}$ and furthermore, by the claim $E(x,r)$ and $E(\xi,\rho)$ are comparable ellipsoids (each is contained in a fixed dilate of the other). Apollonius' theorem implies there are only two possibilities for the circle $C(x,r)$, and we have just seen that (ξ,ρ) must be contained in one of the two $E(x,r)$'s and that they have the proper dimensions. $\qquad\qquad\square$

PROPOSITION 11.22. *For any* $p < \frac{8}{3}$ *there is an estimate*

$$\|M_\delta f\|_q \le C\delta^{-\frac{1}{2}(\frac{3}{p}-1)}\|f\|_p, \quad q = 2p'.$$

This implies, by the proof of Lemma 11.9, that BRK sets have dimension $\geq 2 - \frac{1}{2}(\frac{3}{p} - 1)$ for any $p < \frac{8}{3}$, i.e. dimension $\geq \frac{11}{6}$. Proposition 11.22 was proved (in generalized form) in [36]; it is the partial result which corresponds to the bound (i) in the tangency counting problem. The sharp result ((181) with $p = 3$) incorporating the technique from [19] is proved in [74].

PROOF. This will be similar to the proof of the $\frac{1}{2} + \alpha$ bound in Remark 11.11. The $p = 1$ case is trivial[10] so it suffices to prove the following restricted weak type bound at the endpoint:

$$(191) \qquad \left| \{ r \in [\tfrac{1}{2}, 2] : M_\delta \chi_E(r) > \lambda \} \right| \leq C \left(\frac{|E|}{\delta^{\frac{1}{6}} \lambda^{\frac{8}{3}}} \right)^{\frac{6}{5}}.$$

We may assume in proving (191) that the diameter of the set E is less than one. Consequently in defining $M_\delta f$ we may restrict the point x to the disc $D(0, \frac{1}{4})$. Thus it suffices to prove the following.

Assume that $\lambda \in (0, 1]$ and there are M 3δ-separated values $r_j \in [\frac{1}{2}, 2]$ and points $x_j \in D(0, \frac{1}{4})$ such that $|E \cap C_\delta(x_j, r_j)| \geq \lambda |C_\delta(x_j, r_j)|$. Then

$$(192) \qquad M\delta \leq C \left(\frac{|E|}{\delta^{\frac{1}{6}} \lambda^{\frac{8}{3}}} \right)^{\frac{6}{5}}.$$

We can assume that M is large; for M smaller than any fixed constant (192) holds because $M \neq 0$ implies $|E| \gtrsim \lambda \delta$.

To prove (192) we let μ ("multiplicity") be the smallest number with the following property: there are at least $\frac{M}{2}$ values of j such that

$$(193) \quad \left| E \cap C_\delta(x_j, r_j) \cap \{ x : |\{ i : x \in C_\delta(x_i, r_i) \}| \leq \mu \} \right| \geq \frac{\lambda}{2} \left| C_\delta(x_j, r_j) \right|.$$

The main estimate is

$$(194) \qquad \mu \lesssim M^{\frac{1}{6}} \lambda^{-\frac{5}{3}}.$$

Before proving (194) we introduce some more notation, as follows. For any $t \in [\delta, 1]$ and $\epsilon \in [\delta, 1]$, let

$$a(t, \epsilon) = C_1^{-1} \left(\frac{\delta}{\epsilon} \right)^\alpha \left(\frac{M\delta}{t} + \frac{t}{M\delta} \right)^{-\alpha}.$$

Here α is a sufficiently small positive constant, and C_1 is a positive constant (easily shown to exist) which is large enough that

$$(195) \qquad \sum_{\substack{k \geq 0 \\ l \geq 0}} a(2^k \delta, 2^l \delta) < 1$$

for all M and δ. Let $\overline{\lambda}(t, \epsilon) = a(t, \epsilon)\frac{\lambda}{2}$, $\overline{\mu}(t, \epsilon) = a(t, \epsilon)\mu$, $\overline{M}(t, \epsilon) = a(t, \epsilon)\frac{M}{2}$. Also, for each $i, j \in \{1, \ldots, M\}$ let

$$(196) \qquad \Delta_{ij} = \max\left(\delta, \left| |x_i - x_j| - |r_i - r_j| \right| \right)$$

---

[10]The $p = 2$ case was also known prior to [36]; it follows from results of Pecher [51].

and for each $j \in \{1, \ldots, M\}$, $t \in [\delta, 1]$, $\epsilon \in [\delta, 1]$, let

$$S_{t,\epsilon}(x_j, r_j) \overset{\text{def}}{=} \{i : C_\delta(x_j, r_j) \cap C_\delta(x_i, r_i) \neq \emptyset, \ t \leq |r_i - r_j| \leq 2t$$
$$\text{and } \epsilon \leq \Delta_{ij} \leq 2\epsilon\},$$
$$A_{t,\epsilon}(x_j, r_j) \overset{\text{def}}{=} \{x \in C_\delta(x_j, r_j) : |\{i \in S_{t,\epsilon}(x_j, r_j) : x \in C_\delta(x_i, r_i)\}| \geq \overline{\mu}(t, \epsilon)\}.$$

LEMMA 11.23. *There are numbers $t \in [\delta, 1]$ and $\epsilon \in [\delta, 1]$ with the following property:*

> *There are $\geq \overline{M}(t, \epsilon)$ values of j such that $|A_{t\epsilon}(x_j, r_j)| \geq \overline{\lambda}(t, \epsilon) |C_\delta(x_j, r_j)|$.*

PROOF. This is a routine pigeonhole argument. By the minimality of μ there are at least $\frac{M}{2}$ values of j such that $|\tilde{E}_j| \geq \frac{\lambda}{2}|C_\delta(x_j, r_j)|$ where

$$\tilde{E}_j = E \cap C_\delta(x_j, r_j) \cap \{x : |\{i : x \in C_\delta(x_i, r_i)\}| \geq \mu\}.$$

For any such j and any $x \in \tilde{E}_j$, (195) implies there are $t = 2^k \delta$ and $\epsilon = 2^l \delta$ such that $x \in A_{t\epsilon}(x_j, r_j)$. Consequently, using (195) again, for any such j there are $t = 2^k \delta$ and $\epsilon = 2^l \delta$ such that

$$(197) \qquad |A_{t\epsilon}(x_j, r_j)| \geq \overline{\lambda}(t, \epsilon) |C_\delta(x_j, r_j)|.$$

By (195) once more, there must be a choice of t and ϵ such that (197) holds for at least $\overline{M}(t, \epsilon)$ values of j. This finishes the proof. $\qquad\square$

We fix once and for all a pair (t, ϵ) for which the conclusion of Lemma 11.23 is valid, and will drop the t, ϵ subscripts when convenient, i.e. will denote $\overline{\lambda}(t, \epsilon)$ by $\overline{\lambda}$, etc. We split the proof of (194) into two cases:

(i) $\overline{\lambda} \geq C_2 \sqrt{\frac{\epsilon}{t}}$

(ii) $\overline{\lambda} \leq C_2 \sqrt{\frac{\epsilon}{t}}$

where C_2 is a sufficiently large constant.

In case (i), which is the main case, we let S be the set of M circles in (192), and let $\overline{S}$ be the set of at least $\overline{M}$ circles in Lemma 11.23. Let Q be the set of all quadruples (j, j_1, j_2, j_3) with $C(x_j, r_j) \in \overline{S}$, $C(x_{j_i}, r_{j_i}) \in S$ for $i = 1, 2, 3$ and such that $j_i \in S_{t,\epsilon}(x_j, r_j)$ for each $i \in \{1, 2, 3\}$ and furthermore

$$\text{dist}(C_\delta(x_j, r_j) \cap C_\delta(x_{j_i}, r_{j_i}), C_\delta(x_j, r_j) \cap C_\delta(x_{j_k}, r_{j_k})) \geq C_3^{-1}\overline{\lambda}$$

for all $i, k \in \{1, 2, 3\}$ with $i \neq k$. Here C_3 is a suitable constant which should be chosen before C_2.

We will make two different estimates on the cardinality of Q. On the one hand, the diameter bound in Lemma 11.21 implies that for fixed j_1, j_2, j_3 there are $\lesssim \frac{\epsilon}{\delta}\overline{\lambda}^{-2}$ values of j such that $(j, j_1, j_2, j_3) \in Q$. Also it follows from the definition of Q that there are $\lesssim M \min(M, \frac{t}{\delta})^2$ possible choices for (j_1, j_2, j_3) : there are at most M choices for j_1, and once j_1 is fixed there are $\lesssim \min(M, \frac{t}{\delta})$ possibilities for each of j_2 and j_3, since $|r_{j_1} - r_{j_i}| \leq |r_{j_1} - r_j| + |r_j - r_{j_i}| \leq 4t$ for $i = 2$ or 3. We conclude that

$$(198) \qquad |Q| \lesssim \frac{\epsilon}{\delta} \overline{\lambda}^{-2} M \min\left(M, \frac{t}{\delta}\right)^2.$$

On the other hand, if we fix j with $C(x_j, r_j) \in \overline{S}$ then (provided C_3 has been chosen large enough) we can find three subsets F_1, F_2, F_3 of $A_{t,\epsilon}(x_j, r_j)$ such that $\operatorname{dist}(F_l, F_m) \geq 2C_3^{-1}\overline{\lambda}$, $l \neq m$, and $|F_l| \gtrsim \delta\overline{\lambda}$ for each l. For fixed l, we let S_l be those indices $i \in S_{t,\epsilon}(x_j, r_j)$ such that $F_l \cap C_\delta(x_i, r_i) \neq \emptyset$. The sets $C_\delta(x_i, r_i)$, $i \in S_l$ must cover F_l at least $\overline{\mu}$ times. So

$$\sum_{i \in S_l} |F_l \cap C_\delta(x_i, r_i)| \gtrsim \overline{\mu}\,\overline{\lambda}\,\delta.$$

For each fixed i we have $|F_l \cap C_\delta(x_i, r_i)| \lesssim \frac{\delta^2}{\sqrt{t\epsilon}}$ by Lemma 11.20 (a) Consequently

$$(199) \qquad |S_l| \gtrsim \delta^{-1}\overline{\mu}\,\overline{\lambda}\sqrt{t\epsilon}.$$

It is easy to see using Lemma 11.20 (b) that if $i_l \in S_l$ for $l = 1, 2, 3$ then $(j, i_1, i_2, i_3) \in Q$. So

$$|Q| \gtrsim \overline{M}(\delta^{-1}\overline{\mu}\,\overline{\lambda}\sqrt{t\epsilon})^3.$$

If we compare this with (198) we obtain

$$\overline{\mu}^3 \lesssim \frac{\delta^2}{t^{\frac{3}{2}}\epsilon^{\frac{1}{2}}}\overline{\lambda}^{-5} \min\left(M, \frac{t}{\delta}\right)^2 \frac{M}{\overline{M}},$$

or equivalently

$$\mu^3 \lesssim M^{\frac{1}{2}}\lambda^{-5} \cdot \begin{cases} a(t, \epsilon)^{-9}(\frac{\delta}{\epsilon})^{\frac{1}{2}}(\frac{t}{\delta M})^{\frac{1}{2}} & \text{if } M \geq \frac{t}{\delta}, \\ a(t, \epsilon)^{-9}(\frac{\delta}{\epsilon})^{\frac{1}{2}}(\frac{M\delta}{t})^{\frac{3}{2}} & \text{if } M \leq \frac{t}{\delta}. \end{cases}$$

The expression in the brace is bounded by a constant by the definition of $a(t, \epsilon)$, provided $\alpha < \frac{1}{18}$. So we have proved (194) in case (i).

In case (ii), we fix j with $C(x_j, r_j) \in \overline{S}$ and make the trivial estimate

$$|S_{t\epsilon}(x_j, r_j)| \lesssim \min(M, \tfrac{t}{\delta}).$$

It follows that

$$\overline{\mu}\,\overline{\lambda}\,\delta \lesssim \sum_{i \in S_{t\epsilon}(x_j, r_j)} |C_\delta(x_j, r_j) \cap C_\delta(x_i, r_i)| \lesssim \min(M, \tfrac{t}{\delta})\frac{\delta^2}{\sqrt{t\epsilon}},$$

where we used Lemma 11.20 (a). Thus $\overline{\mu} \lesssim \overline{\lambda}^{-1}\sqrt{\frac{t}{\epsilon}}\min(\frac{M\delta}{t}, 1)$. Using the hypothesis (ii) we therefore have

$$\overline{\mu} \lesssim \overline{\lambda}^{-\frac{5}{3}}\left(\frac{t}{\epsilon}\right)^{\frac{1}{6}}\min\left(\frac{M\delta}{t}, 1\right)$$

i.e.

$$\mu \lesssim \lambda^{-\frac{5}{3}} M^{\frac{1}{6}} \cdot \begin{cases} a(t,\epsilon)^{-\frac{8}{3}} (\frac{\delta}{\epsilon})^{\frac{1}{6}} (\frac{t}{\delta M})^{\frac{1}{6}} & \text{if } M \geq \frac{t}{\delta} \\ a(t,\epsilon)^{-\frac{8}{3}} (\frac{\delta}{\epsilon})^{\frac{1}{6}} (\frac{M\delta}{t})^{\frac{5}{6}} & \text{if } M \leq \frac{t}{\delta} \end{cases}$$

The expression in the brace is bounded by a constant provided $\alpha < \frac{1}{16}$, so we have proved (194).

COMPLETION OF PROOF OF PROPOSITION 11.22   Let $\tilde{E} = \{i : x \in C_\delta(x_i, r_i)| \leq \mu\}$. With notation as above we have

$$|E| \geq |\tilde{E}| \geq \mu^{-1} \sum_j |\tilde{E} \cap C_\delta(x_j, r_j)| \gtrsim \mu^{-1} M \lambda \delta \gtrsim \lambda^{\frac{8}{3}} M^{\frac{5}{6}} \delta$$

by (194). Consequently $(M\delta)^{\frac{5}{6}} \lesssim \frac{|E|}{\delta^{\frac{1}{6}} \lambda^{\frac{8}{3}}}$ and the proposition is proved. $\Box$

Further remarks.

REMARK 11.24. We mention some other recent related work. Schlag [**53**] found an essentially optimal $L^p \to L^q$ estimate in the context of Bourgain's theorem. If we define

$$\mathcal{M}_\delta f(x) = \sup_{1 \leq r \leq 2} \frac{1}{|C_\delta(x, r)|} \int_{C_\delta(x,r)} |f|$$

then there is an estimate

$$\forall \epsilon \ \exists C_\epsilon : \|\mathcal{M}_\delta f\|_5 \lesssim C_\epsilon \delta^{-\epsilon} \|f\|_{\frac{5}{2}}$$

and modulo $\delta^{-\epsilon}$ factors all possible $L^p \to L^q$ bounds for $\mathcal{M}_\delta$ follow by interpolation from this one. Alternate proofs and further related results are in [**56**], [**74**] and [**54**]. On the other hand a number of endpoint questions remain open. The best known is the restricted weak type $(2, 2)$ version of Bourgain's theorem.

REMARK 11.25. A more central open question is the so-called local smoothing conjecture [**58**], [**45**] in $2 + 1$ dimensions. See Section 4 below. This is a problem "with cancellation" and likely not susceptible to purely combinatorial methods without additional input. On the other hand, it would imply (181) with $p = 4$ via the Sobolev embedding theorem and is therefore close to including some of the results of [**19**]. This means perhaps that a proof not involving any combinatorics would have to contain a significant new idea.

REMARK 11.26. One can give a discrete heuristic for the Kakeya problem analogous to the one for the BRK sets problem. What follows is an observation of Schlag and the author.

There is a substantial literature on incidence problems for lines in $\mathbb{R}^3$; these problems appear to be quite difficult and are largely open. One relevant paper is Sharir [**57**], where the following problem is considered:

> Let $\{\ell_j\}_{j=1}^N$ be lines in $\mathbb{R}^3$ and define a *joint* to be a point where three noncoplanar ℓ_j's intersect. Then how many joints can there be?

If $\mathcal{J}$ is the set of joints then as is discussed in [57] the natural conjecture is $|\mathcal{J}| \lesssim N^{\frac{3}{2}}$, which would be sharp by taking $\approx \sqrt{N}$ planes parallel to each of three given planes and considering the lines formed by intersecting two of the planes; any point where three planes intersect will be a joint. The "easy" bound in this problem is $|\mathcal{J}| \lesssim N^{\frac{7}{4}}$ which is proved in [16] using a suitable version of (166). The bound $\forall \epsilon \; \exists C_\epsilon : |\mathcal{J}| \leq C_\epsilon N^{\frac{23}{14}+\epsilon}$ is proved in [57] using similar techniques to [19].

The heuristic is that a bound $|\mathcal{J}| \lesssim N^\alpha$ should imply that (in $\mathbb{R}^3$)

$$\dim(\text{Kakeya}) \geq \frac{\alpha}{\alpha - 1}.$$

Namely, define a μ-fold point in an arrangement of N lines to be a point where at least μ lines intersect with (say) no more than half of these lines belonging to any given 2-plane. Then any bound of the form $|\mathcal{J}| \lesssim N^\alpha$ leads to a corresponding bound $|\mathcal{P}_\mu| \lesssim (\frac{N \log \mu}{\mu})^\alpha$ where $\mathcal{P}_\mu$ is the set of μ-fold points. This may be seen (rigorously) as follows: let $\mathcal{P}_\mu$ be the set of μ- fold points in the arrangement. Let A be a large constant and take a random sample of the N lines according to the following rule: each line belongs to the sample independently and with probability $\frac{A \log \mu}{\mu}$. Then with high probability the sample has cardinality $\lesssim \frac{N \log \mu}{\mu}$. Furthermore, it is not hard to show that any point of $\mathcal{P}_\mu$ will be a joint for the lines in the sample with probability at least $1 - \mu^{-B}$, where B is large if A is large. It follows that with high probability at least half the points of $\mathcal{P}_\mu$ will be joints for the sample, hence $|\mathcal{P}_\mu| \lesssim (\frac{N \log \mu}{\mu})^\alpha$.

Now the heuristic part of the argument: suppose we have a Kakeya set E with (say, Minkowski) dimension β. Fix δ and take a δ-separated set of directions and a line segment in each direction contained in E; this gives an arrangement of $\approx \delta^{-2}$ lines $\{\ell_j\}$. Let E^δ be the δ-neighborhood of E; thus $|E^\delta| \approx \delta^\beta$, so E^δ is made up of roughly $\delta^{-\beta}$ δ-discs. A typical point in the δ-neighborhood of E should belong to roughly $\delta^{-(3-\beta)}$ δ-neighborhoods of ℓ_j's, since otherwise the "low multiplicity" arguments discussed e.g. in Section 2 would show easily that $|E^\delta| \gg \delta^\beta$. Hence if we ignore the distinction between points and δ-discs then we are dealing with an arrangement of δ^{-2} lines with $\delta^{-\beta} \delta^{-(3-\beta)}$-fold points. We conclude that up to logarithmic factors

$$\delta^{-\beta} \lesssim \left(\frac{\delta^{-2}}{\delta^{-(3-\beta)}} \right)^\alpha, \quad \text{i.e.} \quad \beta \geq \frac{\alpha}{\alpha - 1}.$$

Under this heuristic the result of [57] would correspond to an improvement over $\frac{5}{2}$ on Kakeya, and the fact that the joints problem is open would seem to indicate that Questions 1 and 2 are quite difficult even on a combinatorial level, if in fact the answers are affirmative. In this connection, we note that Schlag [55] has proved an analogue of the 3-circle lemma in this context and has used it to give an alternate proof of the result $\dim(\text{Kakeya}) \geq \frac{7}{3}$ (originally due to Bourgain [7]) which corresponds to the result from [16] via $\frac{7}{3} = \frac{7/4}{7/4-1}$. However, it is not easy to put the argument of [57] into the continuum and the author believes that in contrast to the situation considered in [74] it may not be possible to do this in a reasonably straightforward way.

A further remark is that special cases of the three dimensional Kakeya problem correspond to results analogous to [74] with circles replaced by families of curves satisfying the cinematic curvature condition from [58]. For example, the case of sets invariant by rotations around an axis is a problem of this type as is discussed in [36].

11.4. Oscillatory integrals and Kakeya

It seemed appropriate to include a discussion of the basic open problems in harmonic analysis connected with Kakeya, but we will not attempt a complete survey and will not say anything about the proofs of the deeper results. We will just state some well-known open problems and show how they lead to Questions 1 and 2.

Let $\hat{f}$ be the Fourier transform and if m is a given function, then let $T_m f$ be the corresponding multiplier operator,

$$\widehat{T_m f} = m\hat{f}.$$

Two longstanding problems in L^p harmonic analysis are the following:

RESTRICTION PROBLEM Is there an estimate

$$(200) \qquad \|\widehat{fd\sigma}\|_p \lesssim \|f\|_{L^p(d\sigma)}$$

for all $p > \frac{2n}{n-1}$, where σ is the surface measure on the unit sphere $S^{n-1} \subset \mathbb{R}^n$?

BOCHNER–RIESZ PROBLEM Let m_δ be a smooth cutoff to a δ-neighborhood of S^{n-1}, i.e.

$$m_\delta(\xi) = \phi(\delta^{-1}(1 - |\xi|)),$$

where $\phi \in C_0^\infty(\mathbb{R})$ is supported in $(-\frac{1}{2}, \frac{1}{2})$. Then is there an estimate

$$(201) \qquad \forall \epsilon \, \exists C_\epsilon : \|T_{m_\delta} f\|_p \leq \delta^{-\epsilon} \|f\|_p$$

when $p \in [\frac{2n}{n+1}, \frac{2n}{n-1}]$?

Both these problems can be formulated in a number of different ways; the formulations we have given are not the original ones but are well-known to be equivalent to them. In fact it would also be equivalent to prove (200) in the weaker form $\|\widehat{fd\sigma}\|_p \lesssim \|f\|_\infty$, $p > \frac{2n}{n-1}$. This is a consequence of the Stein–Nikisin theory as is pointed out in [7], Section 6.

A third problem of more recent vintage [58] is

LOCAL SMOOTHING Let u be the solution of the initial value problem for the wave equation in n space dimensions,

$$\Box u = 0, \ u(\,\cdot\,,0) = f, \ \frac{\partial u}{\partial t}(\,\cdot\,,0) = 0$$

Then is there an estimate

$$(202) \qquad\qquad \forall \epsilon > 0 \ \exists C_\epsilon : \|u\|_{L^p(\mathbb{R}^n \times [1,2])} \leq C_\epsilon \|f\|_{p,\epsilon}$$

when $p \in [2, \frac{2n}{n-1}]$? Here $\|\cdot\|_{p,\epsilon}$ is the inhomogeneous L^p Sobolev norm with ϵ derivatives.

In all these problems it is well-known that the exponent $\frac{2n}{n-1}$ would be optimal. See [62]. For example, in the last problem this may be seen by considering focussing solutions where f is spread over a δ-neighborhood of the unit sphere and $u(\,\cdot\,,t)$ is mostly concentrated on a δ-disc when $t \in (1, 1 + \delta)$.

When $n = 2$, estimate (200) was proved by Fefferman and Stein and then (201) by Carleson and Sjölin, in the early 1970's (see [62]). Estimate (202) is open even when $n = 2$ however; the known partial results on $L^4(\mathbb{R}^2)$ correspond to loss of $\frac{1}{8}$ derivatives ([45]; an improvement to loss of $\frac{1}{8} - \epsilon$ derivatives appears implicit in [12, p. 60]). In general dimensions, the following implications are known:

$$(202) \Rightarrow (201) \Rightarrow (200) \Rightarrow (165)$$

The first implication is due to Sogge, the second which is deeper is due to Tao [68], and Carbery [15] had shown earlier that the second implication can be reversed in a slightly different context (replace spheres by paraboloids). We refer to [68] for further discussion. Here though we will only be concerned with the last implication which makes the connection with the Kakeya problem. Essentially this is due to Fefferman [25], another relevant reference is [3] and the result as presented here is from [10]. A basic open problem in the area is to what extent the last implication can be reversed. An alternate proof of the two dimensional Carleson–Sjölin result along these lines was given by Córdoba [20]. In three or more dimensions, progress on this problem was initiated by Bourgain (see [10]) who obtained a numerology between partial results which however does not show that (165) would imply (200). For a recent improvement in the numerology see [48] and [72][11].

A problem of a somewhat different nature is

---

[11]*Editor's note:* Wolff [76], [77] made substantial further progress on the problems discussed here, proving a sharp bilinear estimate for the cone and a sharp local smoothing estimate (related to (202)) for a certain range of exponents. Subsequent work inspired by [76], [77] includes [69], [70], [40].

MONTGOMERY'S CONJECTURE Assume $T \leq N^2$. Consider a Dirichlet series

$$D(s) = \sum_{n=1}^{N} a_n n^{is},$$

where $\|\{a_n\}\|_{\ell^\infty} \leq 1$. Let $\mathcal{T}$ be a 1-separated subset of $[0, T]$. Then

$$\forall \epsilon \ \exists C_\epsilon : \sum_{t \in \mathcal{T}} |D(t)|^2 \leq N^\epsilon (N + |\mathcal{T}|) N.$$

An easy consequence (or reformulation) would be that

(203) $$\forall \epsilon \ \exists C_\epsilon : \int_E |D(t)|^2 dt \leq N^\epsilon (N + |E|) N$$

if $E \subset [0, T]$ with the stated hypotheses on T and $D(s)$. This is an estimate on the measure of the set of large values of $D(s)$ and would also imply estimates of L^p norms with $p > 2$. See [9] and e.g. [47] for these remarks as well as some discussion of the relationship between (203) and open problems in analytic number theory. Estimate (203) can perhaps be thought of as an analogue of (200) where the oscillatory sum operator $\{a_n\} \to D(s)$ replaces the extension operator $f \to \widehat{f d\sigma}$. Bourgain [9] showed that (203) is again related to the Kakeya problem.

In the rest of this article, we will discuss implications of this type, i.e.

oscillatory integral estimates $\Rightarrow$ Kakeya estimates

We first show that (200) implies (165), and will record the corresponding implications between partial results. Let us recall the results that would follow from (200) using Hölder's inequality and interpolation with the trivial bound $\|\widehat{f d\sigma}\|_\infty \leq \|f\|_{L^1(d\sigma)}$, say

(204) $$\|\widehat{f d\sigma}\|_q \lesssim \|f\|_{L^p(d\sigma)}, \quad p < \frac{2n}{n-1}, \quad q > \frac{n+1}{n-1} p'.$$

This bound for $p \leq 2$ (plus its endpoint version where $q = \frac{n+1}{n-1}p'$) is a well-known theorem proved by Stein and Tomas in the 1970's and the case $p = q < 2\frac{n+1}{n-1} + \epsilon$ for suitable $\epsilon > 0$ was proved more recently by Bourgain [7] using considerations related to Question 2. See [62] and [10].

PROPOSITION 11.27. *Assume (204) holds for a given $p \geq 2$ and $q \geq 2$. Then, with $r = (\frac{q}{2})'$ and $s = (\frac{p}{2})'$, the restricted weak type (r, s) norm of the Kakeya maximal operator is $\lesssim \delta^{-2(\frac{n}{r}-1)}$. Consequently the Hausdorff dimension of Kakeya sets is $\geq 2r - n = \frac{2q}{q-2} - n$. In particular (200) implies (165).*

PROOF. First let $\{T_j\}_{j=1}^N$, $T_j = T_{e_j}^\delta(a_j)$ be any collection of δ-tubes with δ-separated directions e_j. Let $\tilde{T}_j = \{x \in \mathbb{R}^n : \delta^2 x \in T_j\}$ be the dilation of T_j by a factor δ^{-2}, and let χ_j and $\tilde{\chi}_j$ be the characteristic functions of T_j and $\tilde{T}_j$ respectively. Let C_j be a spherical cap with radius $\approx \delta$ centered at e_j, e.g. $C_j = \{e \in S^{n-1} : e \cdot e_j \geq 1 - C^{-1}\delta^2\}$ where C is a suitable constant.

Take a bump function supported in C_j, say $\phi_j \in C_0^\infty(C_j)$ with $\|\phi_j\|_\infty = 1$, $\phi_j \geq 0$ and $\|\phi_j\|_1 \approx \delta^{n-1}$, and let $\psi_j(\xi) = e^{2\pi i \xi \cdot \delta^{-2} a_j} \phi_j(\xi)$. If $x \in \tilde{T}_j$, then the integral

$$\widehat{\psi_j d\sigma}(x) = \int_{S^{n-1}} \psi_j(\xi) e^{-2\pi i \xi \cdot x} d\xi$$

$$= e^{-2\pi i e_j \cdot (x - \delta^{-2} a_j)} \int_{C_j} \phi_j(\xi) e^{-2\pi i (\xi - e_j) \cdot (x - \delta^{-2} a_j)} d\xi$$

defining $\widehat{\psi_j d\sigma}(x)$ involves no cancellation, so

$$(205) \qquad\qquad |\widehat{\psi_j d\sigma}| \gtrsim \delta^{n-1} \tilde{\chi}_j.$$

Now consider the function $f = \sum_j \epsilon_j \psi_j$ where the ϵ_j are random ± 1's. Since the supports of the ψ_j are disjoint we have

$$\|f\|_{L^p(S^{n-1})} \lesssim (N \delta^{n-1})^{\frac{1}{p}}$$

and therefore, by the assumption (204),

$$(206) \qquad\qquad \|\widehat{f d\sigma}\|_q \lesssim (N \delta^{n-1})^{\frac{1}{p}}$$

for any choice of $\pm$. On the other hand, if we let $\mathbb{E}$ denote expectation with respect to the choices of $\pm$, then by Khinchin's inequality and (205)

$$\mathbb{E}(|\widehat{f d\sigma}|^q) \gtrsim \delta^{q(n-1)} \left(\sum_j \tilde{\chi}_j \right)^{\frac{q}{2}}$$

pointwise. If we integrate this inequality and compare with (206) we obtain

$$\delta^{q(n-1)} \left\| \sum_j \tilde{\chi}_j \right\|_{\frac{q}{2}}^{\frac{q}{2}} \lesssim (N \delta^{n-1})^{\frac{q}{p}}.$$

Rescaling by δ^2, then taking $\frac{q}{2}$th roots,

$$\delta^{2(n-1) - \frac{4n}{q}} \left\| \sum_j \chi_j \right\|_{\frac{q}{2}} \lesssim (N \delta^{n-1})^{\frac{2}{p}}.$$

Now let E be a set, $f = \chi_E$ and

$$\Omega = \{e : f_\delta^*(e) \geq \lambda\}.$$

Let $\{e_j\}_{j=1}^N$ be a maximal δ-separated subset of Ω and for each j choose a δ-tube T_j as above with $|E \cap T_j| \geq \lambda |T_j|$. Then

$$N \lambda \delta^{n-1} \leq \sum_j |T_j \cap E| \lesssim |E|^{1 - \frac{2}{q}} \left\| \sum_j \chi_j \right\|_{\frac{q}{2}} \lesssim |E|^{1 - \frac{2}{q}} (N \delta^{n-1})^{\frac{2}{p}} \delta^{-2(n-1) + \frac{4n}{q}}.$$

Using (169) this implies that

$$|\Omega|^{1 - \frac{2}{p}} \lesssim \lambda^{-1} |E|^{1 - \frac{2}{q}} \delta^{-2(n-1) + \frac{4n}{q}},$$

i.e. $|\Omega|^{\frac{1}{s}} \lesssim \lambda^{-1}|E|^{\frac{1}{r}}\delta^{-2(\frac{n}{r}-1)}$ which is the bound that was claimed. The dimension statement in the proposition then follows from Lemma 11.9, and the last statement also follows by letting $p \to \frac{2n}{n-1}$ and using well-known formal arguments. $\qquad\square$

REMARK 11.28. The original Fefferman construction was of course a counterexample; essentially he showed:

> If the disc multiplier were bounded on L^p with $p \neq 2$, then
> families of tubes with the property in Remark 11.6 could
> not exist.

The paper [3] applies the argument from [25] to the restriction problem in the above way but the result is again formulated as a counterexample. The formulation as an implication concerning the maximal function is from [10].

REMARK 11.29. We present another application of the Fefferman construction which shows the following.

CLAIM. For any $n \geq 2, p > 2, K < \infty$, there are solutions of $\Box u = f$ in n space dimensions, with $\|f\|_\infty \leq 1$, $\operatorname{supp} f \subset D(0,100) \times [0,1]$, and

$$(207) \qquad \int_2^3 \left\|\frac{\partial u}{\partial t}(\cdot,t)\right\|_{L^p(\mathbb{R}^n)} dt > K.$$

The analogous statement with the x-gradient replacing the t-derivative can be proved in a similar way. The statement can be understood as follows: the energy estimate for the wave equation implies via Duhamel's principle that $\|\nabla u(\cdot,t)\|_2 \lesssim \|f\|_2$ if say $t \in (2,3)$ and f is supported in $\mathbb{R}^n \times [0,1]$. The claim says that there can be no such estimate in L^p, $p > 2$, even if one is willing to average in t as in (202) and to restrict to bounded f with compact support. The claim was proved by the author after discussions with S. Klainerman but it is very close to the surface given [25]. The analogous statement for the initial value problem is essentially that (202) fails if the $W^{p\epsilon}$ norm is replaced by the L^p norm on the right hand side; this is a formal consequence of [25] as was probably first observed by Sogge.

The construction below by no means rules out an estimate with loss of ϵ derivatives. In fact the estimate

$$\int_2^3 \left\|\frac{\partial u}{\partial t}(\cdot,t)\right\|_{L^p(\mathbb{R}^n)}^p dt \lesssim \|f\|_{p,\epsilon}^p$$

with $2 < p \leq \frac{2n}{n-1}$ and any $\epsilon > 0$ would follow from (202) via Duhamel.

PROOF OF THE CLAIM. If $x \in \mathbb{R}^n$ then we will use the notation $x = (x_1,\overline{x})$, $\overline{x} \in \mathbb{R}^{n-1}$.

For an appropriate constant C and any small enough δ there is a solution of $\Box u = f$ with

$$\|f\|_\infty \leq 1, \ \operatorname{supp} f \subset \{(x,t) : 0 \leq t \leq 1, \ 0 \leq x_1 \leq 1, \ |\overline{x}| \leq \delta\}$$

and

$$(208) \qquad \left| \frac{\partial u}{\partial t} \right| \geq C^{-1} \quad \text{when } 2 \leq t \leq 3, \, x \in Y^t$$

where Y^t is a subset of $\{x \in \mathbb{R}^n : 2 \leq x_1 \leq 3, \, |\overline{x}| \leq \delta\}$ with measure $\geq C^{-1}\delta^{n-1}$.

This is essentially just the fact that there are high frequency solutions of the wave equation travelling in a single direction tangent to the light cone, which implies we can find f with the indicated support and such that u restricted to $2 \leq t \leq 3$ is also mostly concentrated where $|\overline{x}| \lesssim \delta$. The conclusion then corresponds to conservation of energy.

A rigorous argument can be based on the explicit choice

$$f(x,t) = e^{2\pi i N(x_1 - t)}\phi(x_1)\psi(\delta^{-1}\overline{x})\chi(t)$$

where N is very large, ϕ, ψ, χ are fixed nonnegative C_0^∞ functions, $\psi(0) = 1$, $\mathrm{supp}\,\psi \subset D(0,1)$, $\mathrm{supp}\,\phi = \mathrm{supp}\,\chi = [0,1]$ and ϕ and χ are strictly positive on $(0,1)$. Let u be the corresponding solution of the wave equation. Then u is given by the formula

$$u(x,t) = \int e^{2\pi i x \cdot \xi} \frac{\sin(2\pi(t-s)|\xi|)}{2\pi|\xi|} e^{-2\pi i N s} \hat{\phi}(\xi_1 - N)\delta^{n-1}\hat{\psi}(\delta\overline{\xi})\chi(s) \, d\xi \, ds.$$

One can differentiate for t and then evaluate the resulting integral precisely enough to obtain (208) in the region $|x_1 - t| \leq \frac{1}{2}$, $|\overline{x}| \leq C^{-1}\delta$. We omit the details.

If E is a set in space-time then we will use the notation $E^t = \{x \in \mathbb{R}^n : (x,t) \in E\}$. By Remarks 11.6 and 11.7 we can find disjoint δ-tubes $\{T_j\}_{j=1}^M$ in $\mathbb{R}^n$ ($M \approx \delta^{-(n-1)}$) such that the tubes $\tilde{T}_j$ obtained by translating the T_j's by 2 units along their axes are all contained in a set with small measure $a(\delta)$. Let $\Pi_j = T_j \times [0,1] \subset \mathbb{R}^n \times \mathbb{R}$, and let $\tilde{\Pi}_j = \tilde{T}_j \times [2,3]$. By the first step of the proof there are functions u_j and f_j, $\square u_j = f_j$, with f_j supported on Π_j, $\|f_j\|_\infty \leq 1$, and $\left|\frac{\partial u_j}{\partial t}\right| \geq$ const on a subset $Y_j \subset \tilde{\Pi}_j$ which satisfies $|Y_j^t| \approx \delta^{n-1}$ for each $t \in (2,3)$. Let $Z = \bigcup_j Y_j$; then $|Z^t| \lesssim a(\delta)$ for any $t \in (2,3)$.

Let $\{\epsilon_j\}$ be random ± 1's. Consider the functions $u = \sum_j \epsilon_j u_j$, $f = \sum_j \epsilon_j f_j$, which satisfy $\square u = f$. The Π_j's are disjoint, so $\|f\|_\infty \leq 1$ for any choice of ϵ_j's. On the other hand, by Hölder's and Khinchin's inequalities, for any fixed $t \in (2,3)$ we have

$$\mathbb{E}\left(\int_{Z^t} |\frac{\partial u}{\partial t}(x,t)|^p dx\right)^{\frac{2}{p}} \gtrsim a(\delta)^{-(1-\frac{2}{p})}\mathbb{E}\left(\int_{Z^t} |\frac{\partial u}{\partial t}(x,t)|^2 dx\right)$$

$$= a(\delta)^{-(1-\frac{2}{p})}\int_{Z^t}\sum_j |\frac{\partial u_j}{\partial t}(x,t)|^2 dx$$

$$\gtrsim a(\delta)^{-(1-\frac{2}{p})}\sum_j |Y_j^t| \approx a(\delta)^{-(1-\frac{2}{p})},$$

which shows there can be no estimate of the form

$$\left(\int_2^3 \left\|\frac{\partial u}{\partial t}(\cdot, t)\right\|^2_{L^p(\mathbb{R}^n)} dt\right)^{\frac{1}{2}} \leq C\|f\|_\infty$$

with $p > 2$ when f has support in $D(0, 100) \times [0, 1]$. We then also obtain (207), since an estimate to $L^1_t(L^p_x)$ would imply an estimate to $L^2_t(L^q_x)$ ($\frac{1}{q} = \frac{1}{2}(\frac{1}{2} + \frac{1}{p})$) by interpolation with the energy estimate to $L^\infty_t(L^2_x)$. $\square$

We now discuss the argument from [**9**] relating (203) to (165). Bourgain showed there that Montgomery's conjecture if true would imply Kakeya sets have full dimension and a bound like (165) with a different L^p exponent. We reworked the argument a bit for expository reasons and in order to obtain the precise result (203) $\Rightarrow$ (165).

The logic is that (203) implies a Kakeya type statement for arithmetic progressions, which in turn implies (165) for all n. Thus the implication (203) $\Rightarrow$ (165) follows by combining Propositions 11.30 and 11.31 below.

If $\nu \in (0, 1), \beta \in \mathbb{R}$, then we denote

$$P_\nu^\delta(\beta) = \{x \in [0, 1] : |x - (j\nu + \beta)| < \delta \text{ for some } j \in \mathbb{Z}\}$$

i.e. $P_\nu^\delta(\beta)$ is the δ-neighborhood of the arithmetic progression with modulus ν which contains β, intersected with $[0, 1]$.

PROPOSITION 11.30. *Assume the conjecture (203). Then for any ϵ there is C_ϵ such that the following holds.*

(∗) Fix $\eta \in (0, 1)$, $\delta \in (0, \eta)$. Let $E \subset [0, 1]$ be such that

$$(209) \qquad \forall \nu \in Y \; \exists \beta \in \mathbb{R} : |P_\nu^\delta(\beta) \cap E| \geq \lambda |P_\nu^\delta(\beta)|$$

where $\lambda \in (0, 1]$ satisfies $\lambda \geq C_\epsilon(\frac{\delta^2}{\eta})^{-\epsilon} \cdot \eta$, and where Y is a subset of $(\frac{\eta}{2}, \eta)$ with $|Y| \geq \frac{\eta}{100}$. Then

$$|E| \geq C_\epsilon^{-1}\left(\frac{\delta^2}{\eta}\right)^\epsilon \lambda.$$

PROOF. This will be formally similar to the proof of Proposition 11.27 if one makes the analogy

$$\text{line segment} \longleftrightarrow \text{arithmetic progression}$$

$$\text{spherical cap} \longleftrightarrow \text{interval of integers}$$

CLAIM 1. Let N and T be as in (203) and let ϵ_0 be a suitable constant. Then, for $\nu \in [\frac{N}{2}, N]$ and $\beta \in \mathbb{R}$, the Dirichlet series

$$(210) \qquad d(s) = \sum_{n:|n-[\nu]|\leq \epsilon_0 \frac{N}{\sqrt{T}}} e^{-i\frac{\beta}{[\nu]}(n-[\nu])} n^{is}$$

satisfies

$$|d(s)| \gtrsim \frac{N}{\sqrt{T}}$$

when $s \leq T$ and $\operatorname{dist}(s, 2\pi\nu\mathbb{Z} + \beta) \leq \sqrt{T}$.

PROOF. This is the "short sum" construction in [**9**]. Assume at first that $\nu \in \mathbb{Z}$. The Taylor expansion of the logarithm function shows that

$$n^{is} = \nu^{is} e^{is(\frac{n-\nu}{\nu} + \mathcal{O}(\frac{n-\nu}{\nu})^2))},$$

so that

$$e^{-i\frac{\beta}{\nu}(n-\nu)} n^{is} = \nu^{is} e^{i(n-\nu)) \frac{s-\beta}{\nu} + is\mathcal{O}((\frac{n-\nu}{\nu})^2)}.$$

Thus the sum (210) involves no cancellation and the bound follows immediately. The general case (i.e. $\nu \notin \mathbb{Z}$) follows by replacing ν by $[\nu]$ and noting that this does not significantly affect the hypothesis on s, since if $\operatorname{dist}(s, 2\pi\nu\mathbb{Z} + \beta) \leq \sqrt{T}$ then $\operatorname{dist}(s, 2\pi[\nu]\mathbb{Z} + \beta) \leq \sqrt{T} + C\frac{T}{N} \lesssim \sqrt{T}$. $\quad\square$

We therefore define $\tilde{P}_\nu(\beta) = \{x \in [0, T] : \operatorname{dist}(s, 2\pi\nu\mathbb{Z} + \beta) \leq \sqrt{T}\}$. We also fix a number $\epsilon > 0$ and let C_ϵ be a suitable constant.

CLAIM 2. Assume (203) and let E be a subset of $[0, T]$ with the following property: there is a set $Y \subset [\frac{N}{2}, N]$ with $|Y| \geq \frac{N}{100}$, such that for any $\nu \in Y$ there is $\beta = \beta(\nu) \in \mathbb{R}$ such that $|E \cap \tilde{P}_\nu(\beta)| \geq \lambda |\tilde{P}_\nu(\beta)|$. Then

$$(211) \qquad\qquad |E| \geq C_\epsilon^{-1} N^{-\epsilon} T\lambda$$

provided $\lambda \geq C_\epsilon N^\epsilon \frac{N}{T}$.

PROOF. Let ϵ_0 be as in claim 1, choose a maximal $2\epsilon_0 \frac{N}{\sqrt{T}} + 1$-separated subset $\{\nu_j\}_{j=1}^M \subset Y$, denote $\tilde{P}_j = \tilde{P}_{\nu_j}(\beta_j)$ and let χ_j be the characteristic function of $\tilde{P}_j$. Construct Dirichlet series

$$d_j(s) = \sum_{n:|n-[\nu_j]| \leq \epsilon_0 \frac{N}{\sqrt{T}}} a_n n^{is}$$

via claim 1 so that $|d_j(s)|^2 \gtrsim \frac{N^2}{T} \chi_j$. Let $D(s) = \sum_j \epsilon_j d_j(s)$ where the ϵ_j are random ± 1's. By Khinchin's inequality

$$(212) \qquad\qquad \mathbb{E}(|D(s)|^2) \gtrsim \frac{N^2}{T} \sum_{j=1}^M \chi_j$$

pointwise. On the other hand the coefficient intervals for the d_j are disjoint so for any choice of ± 1, $D(s)$ will be a Dirichlet series with coefficients bounded by 1. Integrating (212) over E and using (203), we obtain

$$\frac{N^2}{T} \sum_{j=1}^M |E \cap \tilde{P}_j| \lesssim \mathbb{E}\left(\int_E |D(s)|^2\right) \lesssim N^\epsilon(N + |E|)\, N.$$

We have $M \approx \sqrt{T}$, and for each j we have $|E \cap \tilde{P}_j| \geq \lambda \frac{T^{\frac{3}{2}}}{N}$. So we obtain $T\lambda \lesssim N^\epsilon(N + |E|)$. Under the stated hypothesis on λ this implies (211). $\quad\square$

Proposition 11.30 follows from Claim 2 by rescaling: set $T = \delta^{-2}$ and $N = \eta\delta^{-2}$, and make the change of variables $x \to Tx$, $\nu \to T\nu$. $\quad\square$

PROPOSITION 11.31. *If* $(*)$ *holds then (165) holds in all dimensions* n.

PROOF. We first observe that $(*)$ implies a generalization of itself via a well-known formal argument (one of the arguments in the Stein–Nikisin theory, see [**60**, p. 146]). Namely, drop the hypothesis $|Y| \geq \frac{\eta}{100}$. Then, with the other hypotheses unchanged,

$$(213) \qquad |E| \gtrsim \lambda \frac{|Y|}{\eta} \left(\frac{\delta^2}{\eta} \right)^\epsilon .$$

To prove (213), let ρY be the dilation of Y by ρ. One can find numbers $\{\rho_j\}_{j=1}^M \subset (\frac{1}{2}, 2)$, where $M \approx \frac{\eta}{|Y|}$, so that $\tilde{Y} \stackrel{\text{def}}{=} \bigcup_j \rho_j Y$ satisfies $|\tilde{Y}| \geq \frac{\eta}{10}$. Let $\tilde{E} = \bigcup_j \rho_j E$. Then $\tilde{E}$ satisfies (209) when $\nu \in \tilde{Y}$ so $|\tilde{E}| \gtrsim \lambda \big(\frac{\delta^2}{\eta}\big)^\epsilon$, hence $|E| \gtrsim \lambda M^{-1} \big(\frac{\delta^2}{\eta}\big)^\epsilon$, which is (213).

Now we consider the Kakeya problem, and will give without detailed proof a few reductions made in [**7**, p. 152].

A. In order to prove (165) it suffices to prove the following inequality: let E be a set in $\mathbb{R}^n$, let Ω be a subset of S^{n-1} with $|\Omega| \geq \frac{1}{2}$, and assume that for any $e \in \Omega$ there is a tube $T_e^\delta(a)$ such that $|T_e^\delta(a) \cap E| \geq \lambda |T_e^\delta(a)|$. Then

$$(214) \qquad \forall \epsilon > 0 \; \exists C_\epsilon : \; |E| \geq C_\epsilon^{-1} \delta^\epsilon \lambda^n .$$

To make this reduction one first observes that (165) is equivalent to the corresponding restricted weak type statement,

$$(215) \qquad \big| \{ e \in S^{n-1} : f_\delta^*(e) \geq \lambda \} \big| \lesssim \delta^{-\epsilon} \frac{|E|}{\lambda^n} ,$$

where $f = \chi_E$, and then uses the above argument from [**60**] to reduce (215) to the case where the left hand side is $\geq \frac{1}{2}$. Furthermore, if $|E \cap T_e^\delta(a)| \geq \lambda |T_e^\delta(a)|$ even for one choice of e and a then clearly $|E| \gtrsim \lambda \delta^{n-1}$. It follows that in proving (214) we can assume $\lambda \geq \delta$.

B. We define $\mathcal{Q}$ to be the unit cube $[0,1) \times \cdots \times [0,1)$. Let N be an integer to be fixed below, such that $\frac{1}{N} < \delta$. If $\nu \in \mathbb{Z}^n$, then we define Q_ν to be the cube $\big[\frac{\nu_1}{N}, \frac{\nu_1+1}{N}\big) \times \cdots \times \big[\frac{\nu_n}{N}, \frac{\nu_n+1}{N}\big)$. When we refer below to a $\frac{1}{N}$-cube we always mean a cube which is of the form Q_ν for some $\nu \in \mathbb{Z}^n$. In proving (214) we can assume that E is contained in $\mathcal{Q}$; this follows easily since the tubes $T_e^\delta(a)$ have diameter $\lesssim 1$. Furthermore we can assume that E is a union of $\frac{1}{N}$-cubes; see [**7**].

C. It is easy to see that $f_\delta^*(e') \leq C f_\delta^*(e)$ if $|e - e'| \leq \delta$, since any tube $T_{e'}^\delta(b)$ can be covered by a bounded number of tubes of the form $T_e^\delta(a)$. Accordingly if Ω is as in A., C_1 is a constant, and if $\mathrm{dist}(e, \Omega) \leq C_1 \delta$ then there is a such that $|T_e^\delta(a) \cap E| \geq C^{-1} \lambda |T_e^\delta(a)|$ where C depends on C_1.

In proving (214) we may assume that

$$\big| \Omega \cap \{ e \in S^{n-1} : e_1 \geq \tfrac{1}{2} \} \big|$$

is bounded below by a constant depending on n only, since we can always achieve this by an appropriate choice of coordinates. In addition, as indicated above we may assume $\lambda \geq \delta$, and we may certainly assume that ϵ is small. Fix integers N and B satisfying the following relations:

$$(216) \qquad B^{-1} N^{2n\epsilon} \approx \lambda \quad \text{and} \quad \frac{B}{N} \approx \delta.$$

Then $N \approx (\delta\lambda)^{\frac{-1}{1-2n\epsilon}}$, so that

$$(217) \qquad N\delta \text{ is large,} \quad N \leq \delta^{-3}, \quad B \text{ is large, and} \quad BN^{-n} \leq B^{-1}.$$

Define a map $\Phi : \mathbb{R}^n \to \mathbb{R}$ via

$$\Phi(x) = \frac{[Nx_1]}{N} + \frac{[Nx_2]}{N^2} + \cdots + \frac{[Nx_{n-1}]}{N^{n-1}} + \frac{Nx_n}{N^n}.$$

Then Φ maps $\mathcal{Q}$ into $[0,1)$. We make a few additional remarks about the definition:

(i) Note the distinguished role played by the last coordinate.

(ii) Φ maps $\frac{1}{N}$-cubes on intervals of length N^{-n}, hence if E is a union of $\frac{1}{N}$-cubes then $|\Phi(E)| = |E|$.

(iii) Suppose that $x \in \mathbb{R}^n$. Then x belongs to a unique $\frac{1}{N}$-cube Q_ν. Define $\tau(x)$ ("tower over x") via

$$\tau(x) = \bigcup (Q_\mu : \mu_j = \nu_j \text{ when } j < n \text{ and } |\mu_n - \nu_n| \leq B).$$

Then, for any x, Φ maps $\tau(x)$ on an interval of length $\frac{2B+1}{N^n}$.

(iv) Suppose that $w = (\frac{k_1}{N}, \ldots, \frac{k_n}{N})$ where the $\{k_j\}$ are *integers*. Set $\nu(w) = \sum_j \frac{k_j}{N^j}$. Then Φ maps any arithmetic progression $\{x + jw\}_{j \in \mathbb{Z}}$ to an arithmetic progression in $\mathbb{R}$ with modulus $\nu(w)$.

A *lattice vector* will be by definition a vector in $\mathbb{R}^n$ of the form

$$w = \left(\frac{k_1}{N}, \ldots, \frac{k_n}{N}\right),$$

where the $\{k_j\}$ are integers with $k_1 \in (\frac{N}{2B}, \frac{N}{B})$ and $\sqrt{\sum_j k_j^2} \leq 2k_1$. Thus any lattice vector w satisfies $|w| \approx \frac{1}{B}$. We note that if $e \in S^{n-1}$ satisfies $e_1 \geq \frac{1}{2}$ then $|e - \frac{w}{|w|}| \lesssim \delta$ for approximately $\frac{N}{B}$ lattice vectors w, namely all the lattice vectors $w = \frac{k}{N}$ which correspond to integer vectors k such that $|k - te| \lesssim 1$ for some t with $t \approx \frac{N}{B}$. Accordingly, for an appropriate constant A there are $\gtrsim (\frac{N}{B})^n$ lattice vectors w such that $\text{dist}(\frac{w}{|w|}, \Omega) \leq A\delta$. We denote this set of lattice vectors by Λ.

If $w \in \Lambda$, then we will abuse our notation slightly and denote the tube $T^\delta_{\frac{w}{|w|}}(a)$ by $T^\delta_w(a)$. By C. above, for each $w \in \Lambda$ we can choose $a \in \mathbb{R}^n$ so that $|T_w^\delta(a) \cap E| \gtrsim \lambda |T_w^\delta(a)|$. It then follows by an averaging argument[12] that there is $a' \in \mathbb{R}^n$ such that

$$(218) \qquad \left| E \cap \left(\bigcup_{j=1}^{B} \tau(a' + jw) \right) \right| \gtrsim \lambda \left| \bigcup_{j=1}^{B} \tau(a' + jw) \right|$$

Now set $\rho = \frac{B}{4N^n}$. By (iv) above, the image of the progression a', $a' + w, \ldots, a' + Bw$ under Φ is an arithmetic progression $\beta, \beta + \nu(w), \ldots, \beta + B\nu(w)$. By (iii), $\Phi(\bigcup_{j=1}^{B} \tau(a' + jw))$ is a union of intervals containing the points of this progression, with the length of each interval being less than ρ and comparable to ρ. Since E and $\bigcup_{j=1}^{B} \tau(a' + jw)$ are unions of $\frac{1}{N}$-cubes, (218) and (ii) then imply that $|\Phi(E) \cap P_\nu^\rho(\beta)| \gtrsim \lambda |P_\nu^\rho(\beta)|$. We conclude:

If $\nu = \nu(w)$ for some $w \in \Lambda$, then there is β such that

$$(219) \qquad |P_\nu^\rho(\beta) \cap E| \gtrsim \lambda |P_\nu^\rho(\beta)|$$

Let $Y = \{\nu \in \mathbb{R} : |\nu - \nu(w)| \leq N^{-n}$ for some $w \in \Lambda\}$. It follows easily that (219) continues to hold (for suitable β) for any $\nu \in Y$. Note that $Y \subset (\frac{1}{2B}, \frac{2}{B})$ (because of the requirement $\frac{N}{2B} \leq k_1 \leq \frac{N}{B}$) and also $|Y| \gtrsim B^{-n}$, since the set $\{\nu(w) : w \in \Lambda\}$ is N^{-n}-separated and has cardinality $\gtrsim (\frac{N}{B})^n$.

Now λ is large compared with $B^{-1} \cdot (B(BN^{-n})^2)^{-\epsilon}$ by (216), (217), so we can apply (213) with $\eta = B^{-1}$, $\delta = BN^{-n}$, and with $\frac{|Y|}{\eta} \gtrsim B^{-(n-1)}$. We conclude that $|\Phi(E)| \gtrsim \lambda B^{-(n-1)}(B^3 N^{-2n})^\epsilon$. Again using (216) and (217), we obtain $|\Phi(E)| \gtrsim \lambda^n N^{-2n^2\epsilon} \geq \lambda^n \delta^{6n^2\epsilon}$. But E is a union of $\frac{1}{N}$-cubes so $|E| = |\Phi(E)|$, and since ϵ is arbitrary this proves (214). $\qquad \square$

---

[12]Namely: let m be the measure of the set $\bigcup_{j=1}^{B} \tau(a' + jw)$; m is clearly independent of a', and furthermore if $x \in \mathbb{R}^n$ is given then the measure of the set $\sigma_x = \{a' : x \in \bigcup_{j=1}^{B} \tau(a' + jw)\}|$ is also comparable to m. If $x \in T_w^\delta(a)$ then, since $\frac{B}{N} \lesssim \delta$ and $|w| \lesssim \frac{1}{B}$, the set σ_x will be contained in $\tilde{T}_w^\delta(a)$, the dilation of $T_w^\delta(a)$ by a suitable fixed constant. It follows that

$$\int_{\tilde{T}_w^\delta(a)} |E \cap (\bigcup_{j=1}^{B} \tau(a' + jw))| \, da' \geq \int_{T_w^\delta(a) \cap E} |\sigma_x| \, dx \geq \lambda m |T_w^\delta(a)| \approx \lambda m |\tilde{T}_w^\delta(a)|,$$

so (218) holds for suitable $a' \in \tilde{T}_w^\delta(a)$.

Bibliography

[1] D. Alvarez, *Bounds for Some Kakeya-type Maximal Functions*, Ph.D. thesis, University of California at Berkeley, 1997.

[2] J. Barrionuevo, *A note on the Kakeya maximal operator*, Math. Res. Lett. 3 (1996), 61–65.

[3] W. Beckner, A. Carbery, S. Semmes, F. Soria, *A note on restriction of the Fourier transform to spheres*, Bull. London Math. Soc. 21 (1989), 394–398.

[4] A.S. Besicovitch, R. Radó, *A plane set of measure zero containing circumferences of every radius*, J. London Math. Soc. 43 (1968), 717–719.

[5] B. Bollobás, *Graph Theory: an introductory course*, Graduate Texts in Mathematics vol. 63, Springer-Verlag, 1979.

[6] J. Bourgain, *Averages in the plane over convex curves and maximal operators*, J. Anal. Math. 47 (1986), 69–85.

[7] J. Bourgain, *Besicovitch type maximal operators and applications to Fourier analysis*, Geom. Funct. Anal. 1 (1991), 147–187.

[8] J. Bourgain, L^p *estimates for oscillatory integrals in several variables*, Geom. Funct. Anal. 1 (1991), 321–374.

[9] J. Bourgain, *On the distribution of Dirichlet sums*, J. Anal. Math. 60 (1993), 21–32.

[10] J. Bourgain, *Some new estimates for oscillatory integrals*, in *Essays on Fourier analysis in honor of Elias M. Stein*, ed. C. Fefferman, R. Fefferman, S. Wainger, Princeton University Press, 1994.

[11] J. Bourgain, *Hausdorff dimension and distance sets*, Israel J. Math 87 (1994), 193–201.

[12] J. Bourgain, *Estimates for cone multipliers*, Operator Theory: Advances and Applications, 77 (1995), 41–60.

[13] J. Bourgain, *On the dimension of Kakeya sets and related maximal inequalities*, Geom. Funct. Anal. 9 (1999), 256–282.[*13]

[14] J. Bourgain, N. Katz, T. Tao, *A sum-product estimate in finite fields and applications*, Geom. Funct. Anal., to appear.[*]

[15] A. Carbery, *Restriction implies Bochner–Riesz for paraboloids*, Proc. Cambridge Phil. Soc. 111 (1992), 525–529.

[16] B. Chazelle, H. Edelsbrunner, L. J. Guibas, R. Pollack, R. Seidel, M. Sharir, J. Snoeyink, *Counting and cutting cycles of lines and rods in space*, Comput. Geom. Theory Appls. 1 (1992), 305- 323.

[17] M. Christ, *Estimates for the k- plane transform*, Indiana Univ. Math. J. 33 (1984), 891–910.

[18] M. Christ, J. Duoandikoetxea, J. L. Rubio de Francia, *Maximal operators related to the Radon transform and the Calderón–Zygmund method of rotations*, Duke Math. J. 53 (1986),189–209.

[19] K. L. Clarkson, H. Edelsbrunner, L. J. Guibas, M. Sharir, E. Welzl, *Combinatorial complexity bounds for arrangements of curves or spheres*, Discrete Comput. Geom. 5 (1990), 99–160.

---

[13]The references indicated by an asterisk were added by the editor.

[20] A. Córdoba, *The Kakeya maximal function and spherical summation multipliers*, Amer. J. Math. 99 (1977), 1–22.

[21] R. O. Davies, *Some remarks on the Kakeya problem*, Proc. Cambridge Phil. Soc. 69 (1971), 417–421.

[22] S. Drury, L^p *estimates for the x-ray transform*, Ill. J. Math. 27 (1983), 125–129.

[23] K. J. Falconer, *The geometry of fractal sets*, Cambridge University Press, 1985.

[24] K. J. Falconer, *On the Hausdorff dimension of distance sets*, Mathematika 32 (1985), 206–212.

[25] C. Fefferman, *The multiplier problem for the ball*, Ann. Math. 94 (1971), 330–336.

[26] R. L. Graham, B. L. Rothschild, J. H. Spencer, *Ramsey Theory*, 2nd edition, Wiley-Interscience, 1990.

[27] M. de Guzman, *Real variable methods in Fourier Analysis*, North-Holland, 1981.

[28] N. Katz, *A counterexample for maximal operators over a Cantor set of directions*, Math. Res. Lett. 3 (1996), 527–536.

[29] N. Katz, *Remarks on maximal operators over arbitrary sets of directions*, Bull. London Math. Soc. 31 (1999), no. 6, 700–710.

[30] N. Katz, *Maximal operators over arbitrary sets of directions*, Duke Math. J. 97 (1999), no. 1, 67–79.[*]

[31] N. Katz, I. Laba, T. Tao, *An improved bound on the Minkowski dimension of Besicovitch sets in R^3*, Ann. of Math. 152 (2000), 383–446.[*]

[32] N. Katz, T. Tao, *Bounds on arithmetic projections and applications to the Kakeya conjecture*, Math. Res. Lett. 6 (1999), 625–630.[*]

[33] N. Katz, T. Tao, *New bounds for Kakeya sets*, J. Anal. Math. 87 (2002), 231–263.[*]

[34] U. Keich, *On L^p bounds for Kakeya maximal functions and the Minkowski dimension in R^2*, Bull. London Math. Soc. 31 (1999), no. 2, 213–221.

[35] J. R. Kinney, *A thin set of circles*, Amer. Math. Monthly 75 (1968), 1077–1081.

[36] L. Kolasa, T. Wolff, *On some variants of the Kakeya problem*, Pacific J. Math. 190 (1999), no. 1, 111–154.

[37] L. Kuipers, H. Niederreiter, *Uniform Distribution of Sequences*, Wiley-Interscience, 1974.

[38] I. Laba, T. Tao, *An X-ray estimate in R^n*, Rev. Mat. Iberoamericana 17 (2001), 375–407.[*]

[39] I. Laba, T. Tao, *An improved bound for the Minkowski dimension of Besicovitch sets in medium dimension*, Geom. Funct. Anal. 11 (2001), 773–806.[*]

[40] I. Laba, T. Wolff, *A local smoothing estimate in higher dimensions*, J. Anal. Math. 88 (2002), 149–171.[*]

[41] J. M. Marstrand, *Packing planes in $\mathbb{R}^3$*, Mathematika 26 (1979), 180–183.

[42] J. M. Marstrand, *Packing circles in the plane*, Proc. London Math. Soc. 55 (1987), 37–58.

[43] W. Minicozzi, C. Sogge, *Negative results for Nikodym maximal functions and related oscillatory integrals in curved space*, Math. Res. Lett. 4 (1997), no. 2-3, 221–237.

[44] T. Mitsis, $(n, 2)$-*sets have full Hausdorff dimension*, preprint, 2001.[*]

[45] G. Mockenhaupt, A. Seeger, C. Sogge, *Wave front sets and Bourgain's circular maximal theorem*, Ann. Math. 134 (1992), 207–218.

[46] G. Mockenhaupt, T. Tao, *Restriction and Kakeya phenomena for finite fields*, Duke Math. J., to appear.[*]

[47] H. L. Montgomery, *Ten lectures on the interface between analytic number theory and harmonic analysis*, CBMS Regional Conference Series in Mathematics, vol. 84, American Mathematical Society, 1994.

[48] A. Moyua, A. Vargas, L. Vega, *Schrodinger maximal functions and restriction properties of the Fourier transform*, Internat. Math. Res. Notices no. 16 (1996), 793–815.

[49] D. M. Oberlin, E. M. Stein, *Mapping properties of the Radon transform*, Indiana Univ. Math. J. 31 (1982), 641–650.

[50] J. Pach, P. Agarwal, *Combinatorial Geometry*, Wiley-Interscience, 1995.

[51] H. Pecher, *Nonlinear small data scattering for the wave and Klein–Gordon equation*, Math. Z. 185 (1984), 261–270.

[52] E. Sawyer, *Families of plane curves having translates in a set of measure zero*, Mathematika 34 (1987), 69–76.

[53] W. Schlag, *A generalization of Bourgain's circular maximal theorem*, J. Amer. Math. Soc. 10 (1997), 103–122.

[54] W. Schlag, *A geometric proof of the circular maximal theorem*, Duke Math. J. 93 (1998), no. 3, 505–533.

[55] W. Schlag, *A geometric inequality with applications to the Kakeya problem in three dimensions*, Geom. Funct. Anal. 8 (1998), no. 3, 606–625.

[56] W. Schlag, C. Sogge, *Local smoothing estimates related to the circular maximal theorem*, Math. Res. Lett. 4 (1997), 1–15.

[57] M. Sharir, *On joints in arrangements of lines in space*, J. Comb. Theory A 67 (1994), 89–99.

[58] C. Sogge, *Propagation of singularities and maximal functions in the plane*, Inv. Math. 104 (1991), 349–376.

[59] C. Sogge, *Concerning Nikodym-type sets in 3-dimensional curved space*, J. Amer. Math. Soc. 12 (1999), no. 1, 1–31.

[60] E. M. Stein, *On limits of sequences of operators*, Ann. of Math. 74 (1961), 140–170.

[61] E. M. Stein, *Maximal functions: spherical means*, Proc. Nat. Acad. Sci. USA 73 (1976), 2174–2175.

[62] E. M. Stein, *Harmonic Analysis*, Princeton University Press, 1993.

[63] E. M. Stein, G. L. Weiss, *Introduction to Fourier analysis on Euclidean spaces*, Princeton University Press, 1971.

[64] E. M. Stein, N. J. Weiss, *On the convergence of Poisson integrals*, Trans. Amer. Math. Soc. 140 (1969), 34–54.

[65] L. Szekely, *Crossing numbers and hard Erdős problems in discrete geometry*, Comb. Prob. Comput. 6 (1997), 353–358.

[66] E. Szemerédi, W. T. Trotter Jr., *Extremal problems in discrete geometry*, Combinatorica 3 (1983), 381–392.

[67] M. Talagrand, *Sur la measure de la projection d'un compact et certaines familles de cercles*, Bull. Sci. Math. (2) 104 (1980), 225–231.

[68] T. Tao, *The Bochner–Riesz conjecture implies the restriction conjecture*, Duke Math. J. 96 (1999), no. 2, 363–375.

[69] T. Tao, *Endpoint bilinear restriction theorem for the cone, and some sharp null form estimates*, Math. Z. 238 (2001), 215–268.*

[70] T. Tao, *A sharp bilinear restriction estimate for paraboloids*, Geom. Funct. Anal., to appear.*

[71] T. Tao, *A new bound for finite field Besicovitch sets in four dimensions*, Pacific J. Math., to appear.*

[72] T. Tao, A. Vargas, L. Vega, *A bilinear approach to the restriction and Kakeya conjectures*, J. Amer. Math. Soc. 11 (1998), no. 4, 967–1000.

[73] T. Wolff, *An improved bound for Kakeya type maximal functions*, Rev. Mat. Iberoamericana 11 (1995), 651–674.

[74] T. Wolff, *A Kakeya type problem for circles*, Amer. J. Math. 119 (1997), 985–1026.

[75] T. Wolff, *A mixed norm estimate for the x-ray transform*, Rev. Mat. Iberoamericana 14 (1998), no. 3, 561–600.

[76] T. Wolff, *Local smoothing type estimates on L^p for large p*, Geom. Funct. Anal. 10 (2000), 1237–1288.*

[77] T. Wolff, *A sharp bilinear cone restriction estimate*, Ann. of Math. 153 (2001), 661–698.*

Historical Notes

For approximately the last seven years Wolff's work had mainly focused on the Kakeya problem and its ramifications in harmonic analysis. Recall that a Kakeya or Besicovitch set in $\mathbb{R}^n$ is a compact set that contains a unit line segment in every direction. It is a classical theorem of Besicovitch that there exist such sets with measure zero. However, in dimension greater than two it is not known whether the HAUSDORFF dimension of such sets needs to be equal to that of the ambient space. From the ground-breaking work of Charles Fefferman in the early 1970s, we now know that this problem lies at the heart of certain questions concerning restriction of the Fourier transform to curved surfaces as well as properties of multipliers with singularities on curved surfaces.

In a well-known paper from 1994 [**H1**], Wolff showed that the Hausdorff dimension of Kakeya sets in $\mathbb{R}^n$ was at least $\frac{n+2}{2}$. He then considered variants of the Kakeya problem with circles in the plane. He had been led to this variant by considering the special case of Kakeya sets $E \subset \mathbb{R}^3$ with cylindrical symmetry around one of the axes. In that case E contains the surface of revolution generated by a line, which is either a hyperboloid of one sheet or a cone. It then suffices to consider the intersection of E with a coordinate plane through the axis of symmetry which now contains a one-parameter family of hyperbolas. The eccentricity of the hyperbolas plays the role of the parameter. The Kakeya problem now takes the following form: Given a set $F \subset \mathbb{R}^2$ so that for any $1 < r < 2$ the set F contains an arc of a hyperbola of length one, say, with eccentricity r, does F have dimension two? It is easy to see that one can replace hyperbolas with circles, in which case r is simply the radius, without changing the problem. It was shown by Besicovitch, Rado, and Kinney that there are such sets F of measure zero, so that the question about dimension is meaningful. In the paper [**H2**], Kolasa and Wolff obtained the lower bound of $\frac{11}{6}$ for the dimension of such sets.

It turns out that the main issue in this problem is to control the number of possible tangencies between circles in a large collection of circles with distinct radii. Kolasa and Wolff controlled the number of these tangencies by means of a combinatorial device that allowed them to pass from a local obstruction to having many tangencies to a global bound on the number of tangencies. The local obstruction on the number of tangencies are the "circles of Appolonius": Given three circles in the plane that are not internally tangent at a single point, there are at most two circles that are tangent to each of the three given ones CIRCLES. The combinatorics involved is known as the Zarankiewicz problem: Given an $N \times N$ matrix with entries 0 or 1 so that there is no 3×3 submatrix containing only 1s, the total number of 1s is at most $N^{\frac{5}{3}}$. Since Kolasa and Wolff only obtained $\frac{11}{6}$ in this way, there is an inherent loss in the passage from the local obstruction to the global bound. This loss was overcome by Wolff in the paper [**H4**] by adding the technique of cell decomposition from [**H3**] to his approach and he thus achieved the optimal lower bound of two for the dimension of Besicovitch-Rado-Kinney sets in the plane. Shortly thereafter, Wolff obtained a generalization of his Kakeya maximal function bound of $\frac{n+2}{2}$ by allowing for parallel tubes. This is known as the X-ray transform, see [**H6**], and is of great importance in a variety of problems.

In more recent work, Wolff obtained an improvement in the Falconer distance set problem. In [**H9**], he showed that a set $E \subset \mathbb{R}^2$ of dimension bigger than $\frac{4}{3}$ has a distance set of positive length. Jean Bourgain had previously shown that sets with dimension bigger than $\frac{13}{9}$ have this property. It is conjectured that the bound of $\frac{4}{3}$ can be lowered even further. Wolff obtained his bound by establishing the best possible decay rate on circular means of Fourier transforms of measures, a problem posed by Mattila. This latter result motivated the recent important development [**H10**], where the sharp restriction bound of the Fourier transform to the cone in $\mathbb{R}^4$ is obtained. This is the first example of a surface with two nonvanishing principal curvatures where a sharp restriction bound on the Fourier transform has been proved. Another surface of interest is the sphere in $\mathbb{R}^3$, where the restriction conjecture (due to Elias Stein) is open. It is known that the restriction conjecture for the sphere implies the Kakeya conjecture, see e.g. [**H7**]. Finally, Wolff combined the methods from [**H4**] and [**H10**] to obtain a sharp local smoothing bound for the wave equation in a certain range of exponents, see [**H11**].

Concurrently with his main program that we have just outlined, Wolff was applying harmonic analysis techniques to other areas, such as mathematical physics. In [**H5**] Shubin, Vakilian, and Wolff studied the Anderson Bernoulli model on the line by means of some refined uncertainty principle ideas. It is well known that under suitable conditions on sets $E \subset \mathbb{R}^n$ and $F \subset \mathbb{R}^n$ there is a constant C such that

$$\|f\|_{L^2(\mathbb{R}^n)} \leq C(\|f\|_{L^2(E^c)} + \|\hat{f}\|_{L^2(F^c)}) \qquad \forall\, f \in L^2(\mathbb{R}^n).$$

More precisely, Amrein and Berthier showed that this holds if E and F have finite measure. Also, a theorem of Logvinenko and Sereda states that it holds if F is the unit ball, say, and E is a set that is sufficiently thick in the sense that for some R and $\alpha > 0$

$$|E \cap B(x,R)| > \alpha |B(x,R)| \qquad \forall\, x \in \mathbb{R}^n.$$

A new instance of this fact was used in [**H5**] to simplify the approach of Campanino, Klein, Martinelli, and Perez, which is based on the supersymmetric replica method, and thus extend it to the case of Bernoulli potentials. Moreover, they gave an alternative proof of Le Page's theorem on Hölder continuity of the integrated density of states. In doing so they were able to show that the Hölder exponent stays bounded away from zero as the disorder goes to zero, at least away from the band edges. After the seminal work of Furstenberg on products of random matrices, it is natural to study the Anderson model on the line by means of the invariant measure of the Schrödinger cocycle that it gives rise to. Motivated by this connection, Wolff used the methods from [**H5**] to obtain estimates for spectral gaps of certain representations of semisimple Lie groups, see [**H8**].

Very recently, Wolff returned to the Anderson model. With Klopp he obtained novel estimates for Lifshitz tails at the band edges for the integrated density of states in the random model on $\mathbb{R}^2$, see [**H12**]. This used delicate estimates on the norm of oscillatory integral operators with nonconstant analytic phases obtained by Phong and Stein. Finally, in [**H13**] it is shown that the discrete random model with disorder λ on $\mathbb{Z}^2$ has the property that a.s. most eigenfunctions have Fourier transforms localized to annuli of thickness $\lambda^{2-\epsilon}$ for any $\epsilon > 0$. By the uncertainty principle this then implies that the localization length of most eigenfunctions is at least $\lambda^{-2+\epsilon}$.

Bibliography

[H1] T. Wolff, *An improved bound for Kakeya type maximal functions.* Rev. Mat. Iberoamericana 11 (1995), no. 3, 651–674.

[H2] L. Kolasa, T. Wolff, *On some variants of the Kakeya problem.* Pacific J. Math. 190 (1999), no. 1, 111–154.

[H3] K. L. Clarkson, H. Edelsbrunner, L. J. Guibas, M. Sharir, E. Welzl, *Combinatorial complexity bounds for arrangements of curves and spheres.* Discrete Comput. Geom. 5 (1990), no. 2, 99–160.

[H4] T. Wolff, *A Kakeya-type problem for circles.* Amer. J. Math. 119 (1997), no. 5, 985–1026.

[H5] C. Shubin, R. Vakilian, T. Wolff, *Some harmonic analysis questions suggested by Anderson-Bernoulli models.* Geom. Funct. Anal. 8 (1998), no. 5, 932–964.

[H6] T. Wolff, *A mixed norm estimate for the X-ray transform.* Rev. Mat. Iberoamericana 14 (1998), no. 3, 561–600.

[H7] T. Wolff, *Recent work connected with the Kakeya problem.* Prospects in Mathematics (Princeton, NJ, 1996), 129–162, Amer. Math. Soc., Providence, RI, 1999.

[H8] T. Wolff, *A general spectral gap property for measures.* J. Anal. Math. 88 (2002), 27-34.

[H9] T. Wolff, *Decay of circular means of Fourier transforms of measures.* Internat. Math. Res. Notices 1999, no. 10, 547–567.

[H10] T. Wolff, *A sharp bilinear cone restriction estimate.* Ann. of Math. 153 (2001), 661–698.

[H11] T. Wolff, *Local smoothing type estimates on L^p for large p.* Geom. Funct. Anal. 10 (2000), 1238–1288.

[H12] F. Klopp, T. Wolff, *Internal Lifshitz tails for random Schrödinger operators.* J. Anal. Math. 88 (2002), 63-148.

[H13] W. Schlag, C. Shubin, T. Wolff, *Frequency concentration and localization lengths for the Anderson model at small disorders.* J. Anal. Math., 88 (2002), 173-220.

Titles in This Series

29 **Thomas H. Wolff (Izabella Łaba and Carol Shubin, editors),** Lectures on harmonic analysis, 2003

28 **Skip Garibaldi, Alexander Merkurjev, and Jean-Pierre Serre,** Cohomological invariants in Galois cohomology, 2003

27 **Sun-Yung A. Chang, Paul C. Yang, Karsten Grove, and Jon G. Wolfson,** Conformal, Riemannian and Lagrangian geometry, The 2000 Barrett Lectures, 2002

26 **Susumu Ariki,** Representations of quantum algebras and combinatorics of Young tableaux, 2002

25 **William T. Ross and Harold S. Shapiro,** Generalized analytic continuation, 2002

24 **Victor M. Buchstaber and Taras E. Panov,** Torus actions and their applications in topology and combinatorics, 2002

23 **Luis Barreira and Yakov B. Pesin,** Lyapunov exponents and smooth ergodic theory, 2002

22 **Yves Meyer,** Oscillating patterns in image processing and nonlinear evolution equations, 2001

21 **Bojko Bakalov and Alexander Kirillov, Jr.,** Lectures on tensor categories and modular functors, 2001

20 **Alison M. Etheridge,** An introduction to superprocesses, 2000

19 **R. A. Minlos,** Introduction to mathematical statistical physics, 2000

18 **Hiraku Nakajima,** Lectures on Hilbert schemes of points on surfaces, 1999

17 **Marcel Berger,** Riemannian geometry during the second half of the twentieth century, 2000

16 **Harish-Chandra,** Admissible invariant distributions on reductive p-adic groups (with notes by Stephen DeBacker and Paul J. Sally, Jr.), 1999

15 **Andrew Mathas,** Iwahori-Hecke algebras and Schur algebras of the symmetric group, 1999

14 **Lars Kadison,** New examples of Frobenius extensions, 1999

13 **Yakov M. Eliashberg and William P. Thurston,** Confoliations, 1998

12 **I. G. Macdonald,** Symmetric functions and orthogonal polynomials, 1998

11 **Lars Gårding,** Some points of analysis and their history, 1997

10 **Victor Kac,** Vertex algebras for beginners, Second Edition, 1998

9 **Stephen Gelbart,** Lectures on the Arthur-Selberg trace formula, 1996

8 **Bernd Sturmfels,** Gröbner bases and convex polytopes, 1996

7 **Andy R. Magid,** Lectures on differential Galois theory, 1994

6 **Dusa McDuff and Dietmar Salamon,** J-holomorphic curves and quantum cohomology, 1994

5 **V. I. Arnold,** Topological invariants of plane curves and caustics, 1994

4 **David M. Goldschmidt,** Group characters, symmetric functions, and the Hecke algebra, 1993

3 **A. N. Varchenko and P. I. Etingof,** Why the boundary of a round drop becomes a curve of order four, 1992

2 **Fritz John,** Nonlinear wave equations, formation of singularities, 1990

1 **Michael H. Freedman and Feng Luo,** Selected applications of geometry to low-dimensional topology, 1989